First Lessons in Beekeeping

Complete and Unabridged

First Lessons in Beekeeping

By C. P. Dadant

Introduction

BY DR. C. C. MILLER.

Among those who were instrumental in introducing advanced methods in bee-culture among the beekeepers of Europe in the last century, especially in overcoming opposition to the movable-frame hive, Charles Dadant stands forth conspicuous. In France, "Dadant" and "Dadant hive" are household words among beekeepers. His great influence was used by means of his facile pen, for most of his life was spent in this country, where he was less known because not so familiar with the English language as with his native tongue, the French.

Yet his greatest legacy was not to France but to America. That legacy was his only son, Camille P. Dadant. Intimately associated with his father from his birth until the close of the long life of thé father, the younger Dadant had a schooling as a beekeeper that can fall to the lot of few.

Mr. Dadant collaborated with his father in revising the great work of Langstroth. He has written much and well not only for the bee journals of this country, but of France as well. His very practical writings as editor of The American Bee Journal are well known.

All this, together with his long and successful career as a bee-keeper gives warrant that the present work shall be a safe and sane guide to those entering upon the fascinating pursuit of beekeeping.

Marengo, Illinois, March 3, 1916.   C. C. MILLER.

Preface

This short treatise for beginners is an entirely rewritten edition of the now exhausted book published in 1911 by George W. York and Co.

Less extensive than our revision of the "Hive and Honey Bee," it yet contains the most practical of modern methods available in our day. But as simplicity is important we have kept it in mind and difficult methods will not be found here. However a still more elementary work is given in our "Bee Primer," which, at the low price of fifteen cents, has found a great welcome among prospective beekeepers.

This book is especially intended for colleges and schools giving short courses in bee culture. A few years ago such courses were not thought of. But they are annually becoming more numerous. Blind beekeeping is still less profitable than blind farming. The hive has long been a sealed book. It should be opened to the prospective apiarist before he attempts to keep bees. The bee owner who depends upon luck is an obstruction to the success of others, for disease and degenerescence of his bees are sure to follow from his lack of knowledge and method.

The different subjects treated in this work are marked, for reference, at the head of the paragraphs, with serial numbers. When reference is made to another part of the book, the serial number is inserted in parenthesis, so that the student will easily find all references to the subject which he studies. Likewise, the index refers to the paragraphs and not to the pages of the book.

The different species of living animals number over a quarter of a million. Among this vast concourse of life, for instructive lessons none can rival the marvelous transformations that insect life undergoes in its development! The repulsive maggot of today may tomorrow be the active little fly, visiting leaf and flower. The repugnant caterpillar may to-morrow be decked with green and gold, through its speedy transformation to the butterfly, of brilliant tints and gorgeous beauty.

This is not more wonderful than are the transformations from the egg to the tiny larva, from the larva to the pupa, and from the pupa to the fully developed honeybee, with its wondrous instincts and marvelous habits. There is a fascination about the apiary that is indescribable. Every scientific beekeeper is an enthusiast. The economy of the beehive presents to the thoughtful student both admiration and delight.

A single bee, with all its energy, collects but a tiny drop of honey at each trip to the field, in the best season, yet the colony to which it belongs may harvest hundreds of pounds of surplus for its owner, in a single year.

In fructifying the flowers, too, bees present us with a field of study. Many plants absolutely require the visits of bees or other insects to disturb their pollen, and thus fertilize them. Hence, Darwin wisely remarks, when speaking of clover and heartsease: "No bees, no seed, no increase of the flower; the more visits from the bees, the more seeds from the flower; the more seeds from the flowers, the more flowers from the seeds." Darwin mentions the following experiment: "Twenty heads of white clover, visited by bees, produced an average of twenty-seven seeds per head; while twenty heads, so protected that bees could not visit them, produced not one seed."

Since the Darwin experiment, hundreds of scientists have made tests of this same subject. Bulletin No. 289 of the United States Department of Agriculture, published September 21, 1915, details at length the experiments made at the Indiana Experiment Station by Messrs Wiancko and Robbins and at the Iowa Experiment Station of Ames, by Messrs Hughes, Pammel and Martin. They confirm Darwin's statements and show that clover can produce only "an occasional seed from self pollination, that the pollen must come from a separate plant in order to effect fertilization". They also show that the honeybee is as efficient a pollinator of red clover as the bumblebee, whenever it is able, by the shortness of the corolla, to work upon it.

Ancient sages, among whom were Homer, Herodotus, Cato, Aristotle, Varro, Virgil, Pliny and Columella, composed poems extolling the activity, skill and economy of bees, and in modern times among such authors have been Swammerdam, a Dutch naturalist; Maraldi, an Italian mathematician and astronomer; Schirach, a Saxon agriculturist; Reaumur, inventor of a thermometor; Butler, who first asserted the existence of a queenbee; Wildman; Della Rocca; Duchet; Bonnet, a Swiss entomologist; Dr. John Hunter; and Francis Huber, who, though totally blind, was noted for his many minute observations, by aid of his assistant, Burnens, which caused quite a revolution in ancient theories concerning honeybees. Nearer to our day, we may mention as the leaders of modern practical apiculture: Dzierzon, Von Siebold, John Lubbock, L. L. Langstroth, Samuel Wagner, M. Quinby, Adam Grimm, J. S. Harbison, Capt. J. E. Hetherington, Prof. A. J. Cook, G. M. Doolittle, Dr. C. C. Miller, A. I. Root and his sons, Chas. Dadant, E. W. Alexander,Thos. Wm. Cowan, Frank R. Cheshire. Edward Bertrand, and a host of others.

It is out of the question to make mention of the students and teachers of 20th Century beekeeping. They are so numerous that a complete list would be irksome.

Hamilton, Illinois, January 15, 1917.   C. P. DADANT.

Contents

The numbers refer to the paragraphs and not to the pages.

Natural History of the Honey-Bee

The Races of Bees

§ 1. Of the different races of the honeybee, the common or black bee is the most numerous, though it is less desirable than the Italian, which was known to the ancients several hundred years before the Christian Era, and is mentioned by Aristotle and Virgil. The Egyptian, Carniolan, Cyprian, Caucasian, and others, have also been tried. But the Italian (123) is the favorite in the United States, because of its activity, docility, prolificness and beauty.

A Colony of Bees

§ 2. In its usual working condition, a colony of bees contains a fertile queen, many thousands of workers (more or less numerous according to the season of the year), and in the busy season from several hundred to a few thousand drones.

The Queen

§ 3. The mother-bee, as she is often called, is the only perfect female in the colony and is the true mother of it. Her only duty is to lay the eggs for the propogation of the species. She is a little larger than the worker but not so large as the drone. Her body is longer than that of the worker, but her wings are proportionately shorter. Her abdomen tapers to a point. She has a sting, but it is curved, and she uses it only upon royalty; that is to say, to fight or destroy other queens—her rivals.

§ 4. The queen usually leaves the hive only when accompanying a swarm. However, she takes a flight when about five or six days old, to mate with a drone, outside, upon the wing. Once fertilized, she is so for life, though she often lives three or four years (30). On her return to the hive, after

Fig. 1—The Queenbee

mating, if she has been fecundated, the male organs may be seen attached to her abdomen.

§ 5. If for some reason the queen is unable to mate within the first three weeks of her life, she loses the desire to mate, but is

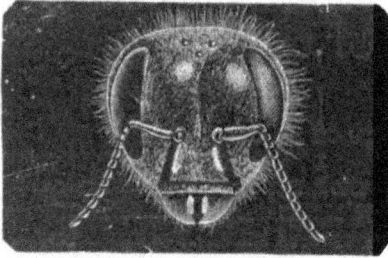

Fig. 2—Head of Queen (magnified.)

nevertheless able to lay eggs that will hatch, as will be shown further (9). These produce only drones. In about two days after mating, she commences to lay, and she is capable, if prolific, of laying three thousand or more eggs per day. These are regularly deposited by her in the cells, within the breeding apartment or body of the hive. When a queen lays eggs in the super or honey receptacle, which is usually provided over the hive-body, it is a sign that the hive is full. Small hives are objectionable because their

Fig. 3—Queen Laying; Surrounded by Workerbees.

limited space often causes the queen to desert the breeding apartment and induce swarming.

§ 6. Instinct teaches the workers the necessity of having a queen that is prolific, and should she become barren from any cause, or be lost or even decrease in her fertility (101-5) during the breeding season, or die (118) from old age or from accident, they immediately prepare to rear another to take her place. This they do by building queencells (Fig. 5.) (34) which they supply with eggs from worker-cells.

The bees also rear queens when preparing to swarm (96); the first queen hatched destroys the others and the bees usually help her to do it unless they wish to swarm (98) again.

§ 7. By feeding the embryo queen with royal jelly, the egg that would have produced a worker had it remained in a worker-cell, becomes a queen.

The name "royal jelly" (33)

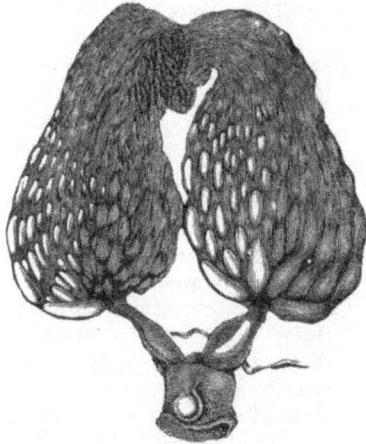

Fig. 4—Ovaries of Queen
(magnified.)

is probably a misnomer, though used by most authors. It seems evident that the royal jelly is the same food which is given to the larva of the worker-bee during the first three days of its existence, but at the end of that time it is changed, for the worker, to a coarser food or pap, while the same jelly in plentiful supply is given to the queen-larva during the entire time of its growth.

§ 8. The ovaries of the queen, occupying a large portion of the abdomen, are two pear-shaped bodies, composed of 160 to 180 minute tubes, the tubes being bound together by enveloping air-vessels. A highly magnified view is here given (Fig. 4.) The germs of the eggs originate in the upper ends of the tubes which compose the ovary, and the eggs develop in their onward passage, so that at the time of the busy laying season each one of the tubes will contain, at its lower end, one or more mature eggs, with several others in a less developed state following them. These tubes terminate on each side in the oviduct, through which the egg passes into the vagina; in the cut, an egg will be seen in the oviduct on the right.

A globular sac will be noted, attached to the main oviduct by short, tubular stem. A French naturalist, M. Audouin, first discovered the true character of this sac as the spermatheca, which contains the male semen; and Prof. Leuckart computes its size as sufficient to contain, probably, twenty-five millions of seminal filaments. It seems hardly possible that so large a number should ever be found in the spermatheca, as it would require nearly twenty years to exhaust the supply, if the queen should lay daily 2000 eggs, 365 days in the year, and each egg be impregnated. Each egg which receives one or more of of the seminal filaments in passing produces a worker or queen, while an unimpregnated egg produces only a drone. The spermatheca of an unfecundated queen contains only a transparent liquid with no seminal filaments, and the eggs of such a queen produce only drones, whether they are laid in large or small cells. The size of the cell has therefore no influence on the sex.

Fig. 5. Queencells.

§ 9. This ability of a queen to lay eggs which hatch into drones, without fertilization, belongs only to a few female insects and is called "parthenogenesis." This was discovered in queenbees by Dzierzon. Whether the queen has been for some cause unable to meet a drone or to fly in search of one, or whether the drone's organs were sterile, or their supply exhausted, or whether yet she has been rendered infertile by refrigeration, in any of these cases a queen may lay eggs which hatch only as drones. Such a queen is, of course, worthless, and should be superseded by the apiarist.

§ 10. The queen usually lays from February to October, but very early in the spring she lays sparingly. When fruit and flowers bloom, and the bees are getting honey and pollen, she lays most rapidly.

The Drones

§ 11. These are non-producers, and live on the toil and industry of others. They are the males, and have no sting—neither have they any means of gathering honey or secreting wax, or doing any work that is even necessary to their own support, or the common good of the colony.

§ 12. The drones are shorter, thicker and more bulky than the queen, and their wings reach the entire length of their body. They are much larger and clumsier than the workers, and like the queen and workers are covered with short but fine hair. Their buzzing when on the wing is much louder and differs from that of the others. Their only use is to serve the queen when on her "bridal trip."

Not more than one in a thousand is ever privileged to perform that duty, but as the queen's life is very valuable, and the dangers surrounding her flight are numerous. it is necessary to have a sufficient number of them, in order that her absence from the hive may not be protracted.

Fig. 6. The Drone.

That is why hundreds and often thousands of drones are reared in each colony during the breeding and swarming season. In domestication, when dozens and sometimes hundreds of colonies are kept in an apiary, the choice colonies alone should be permitted to rear drones in large numbers for reproduction.

§ 13. It is said that some queens need to mate twice before fertilization is fully accomplished. But the average queen mates but once and the drone, in the act of copulation, loses his life, dying instantly.

Fig. 7. Sexual Organs of the Drone.
a. a. testicles; b. b. mucous glands; c, seminal duct; d, formation of

§ 14. After the swarming season is over, or should the honey season prove unfavorable and the crop short, they are mercilessly destroyed by the workers.

Should a colony lose its queen, the drones will be retained later; instinct teaching them that, without the drone, the young queen would remain unfertile, and the colony soon become extinct.

§ 15. When comparing the head of the drone (Fig. 8), with those of the queen and the worker (Figs. 2 and 10), one readily notices the compound eyes, those crescent-shaped projections on each side of the head. They are much larger in the drone than in either of the others, and this is ascribed by scientists to the necessity of finding the queen in the air, on the wing. The facets composing these eyes number some 25,-000 in the head of the drone, so that they can see in all directions. The three small points in a triangle at the top of the head are small eyes or ocelli, which are probably used to see in the dark, within the hive, and at short range.

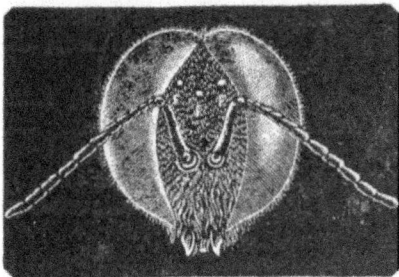

Fig. 8—Head of Drone (magnified.)

§ 16. It has been common among beekeepers to believe that the drones serve another purpose in the hive, aside from their use as males. It is said that they keep the brood warm. As a matter of course, they keep themselves upon the brood combs, when permitted, as much to enjoy the natural warmth of the living grubs as to keep them warm. But the fallacy of the belief in their being required to keep up the warmth is clearly seen when the bees drive them out and destroy them at the least reverse in the temperature. The greatest number of drones are reared in the warmest part of the season and their uselessness for other purposes than the fertilization of the queens is very positively proven.

The Workers

§ 17. These are undeveloped females, and they do all the work that is done in the hive. They secrete the wax, build the comb, ventilate the hive, gather the pollen for the young, and honey for all, feed and rear the brood, and fight all the battles necessary to defend the colony.

Of the three kinds of bees these are the smallest, but constitute the great mass of the population. They possess the whole ruling power of the colony and regulate its economy.

§ 18. The details of the **head** of a bee are very interesting. We have already mentioned, when speaking of the drone, the compound eyes, which are larger and contain a greater number of facets in the male than in either the queen or the worker.

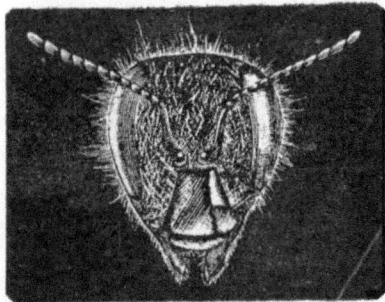

§ 19. The bees have short, thick, smooth **mandibles**, working sidewise instead of up and down as in higher animals. These mandibles have no teeth like those of wasps and hornets, and yet enable them to tear the soft corolla of flowers and to build their combs out of wax. They are therefore incapable of cutting the smooth skin of sound fruits of any kind.

Fig. 9.
The Worker.

§ 20. The **tongue** of the honeybee is made of several parts, ligula, palpi and maxillae (fig. 12). The central part or ligula is grooved like a trough. When at rest it is folded below the mentum or chin.

Fig. 10—Head of Worker.
(magnified)

§ 21. In the head and thorax are three pairs of **salivary glands,** two of which at least are evidently used to produce the saliva which changes the chemical condition of the nectar of blossoms into that of honey. The largest pair of glands is supposed to be used in the production of the pap for the larvae, as will be seen further (33).

§ 22. The **antennæ** or feelers are the two long horns which protrude from the head of the bee. These exist in all insects. The popular name of "feelers" is very proper, for it is with these antennæ that the bee examines every body or thing with which is comes in contact. They appear to serve the purpose of smell, touch and hearing. It is however claimed by a modern scientist, Mr. McIndoo, of the Bureau of Entomology at Washington, that the bees do not smell through the antennæ and that there are organs of smell, located in other parts of the body, at the joints of wings, legs, etc. It is true also that the organs of breathing are not in the head, but in the abdomen, between the rings or segments of the third section of the body. However, until further proof is adduced, we must continue, with all entomological students, to ascribe the detection of the most minute odors to the antennæ, since it is with these organs that they examine the blossoms, the combs, their friends and their enemies. As there are usually tens of thousands of bees in a colony and they very readily recognize their own members, it must be with the antennæ that this recognition is achieved.

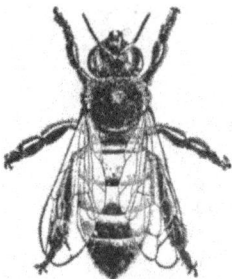

§ 23. The honey-sac (Fig. 13), or first stomach, is located in the abdomen or third segment of the body of the bee. From this stomach, the bee may at will digest a part of the honey, by forcing it to the second stomach for the nourishment of its body, or it may be discharged back through the mouth into the cells for future use (49). Another use of the honey is to make comb, as will be explained further (38).

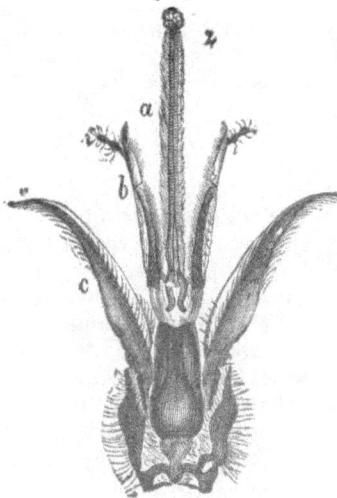

Fig. 11—The Worker
(magnified).

Fig. 12—Tongue and
Appendages.

Fig. 13—The Digestive
Apparatus.

Fig. 12—(a) tongue, (b) labial palpi, (c) maxillæ
Fig. 13—(a) tongue, (b) esophagus, (c) honey sack, (d) digesting stomach, (e) malpighian tubes, (f) small intestine, (g) large intestine.

§ 24. The honeybee has four **wings** and six **legs,** all fastened to the corslet or second segment of the body. The wings, in pairs, fold upon each other to enable them to enter within the cells where the brood is reared and where the honey is stored. In flight the two sections of these wings are braced together by the use of very fine hooks, which enable them to present a greater surface in contact with the air.

§ 25. We will not go into the details of the different segments of the legs of the honey bee. But it is well to say that each leg is supplied at its extremity with claws which permit the bees to hang to each other in clusters. They also have near the claw a small "rubber-like pocket" which secretes a sticky substance. This enables the bee, like the fly and many other insects, to fasten itself and walk with ease upon any smooth surface, such as a pane of glass or a ceiling.

Fig. 14—Anterior Leg of Worker (magnified.)

§ 26. The anterior legs (Fig. 14.) are provided with a notch and a thumb-like spine or "velum" A, B. C. which is used by the insect to cleanse the antenna. The motion made for this purpose is often noticed in house flies as well as in bees.

§ 27 The third pair of legs of the workerbee have a hollow cavity (Fig. 15 A A), called the pollen basket, which enables it to carry home the pollen of flowers, which some people, when they see them so loaded, imagine to be wax, but which is used to make the pap or jelly for the young. This pollen (55) is popularly called bee-bread and is the fertilizing dust of flowers.

It is peculiar and wonderful that neither the queen nor the drones are supplied with these pollen baskets (Fig. 16). They would have no use for them since they never work in the field.

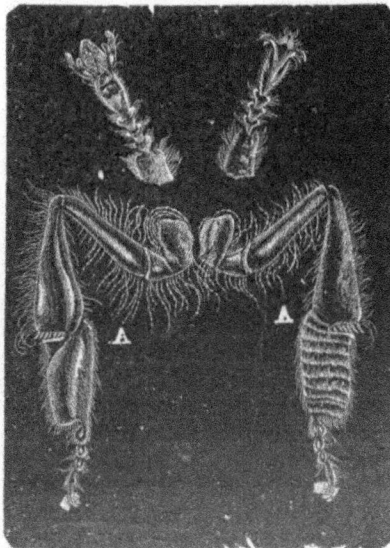

Fig. 15—Posterior Legs of Worker (magnified.) A A Pollen-Baskets.

§ 28. The **ovaries,** or egg pouches, which are very large in the queen, are almost absent in the workers, who are therefore incomplete females (121) and unfit for mating, although they may occasionally be able to lay a few eggs which hatch as drones.

Fig. 16—Posterior Legs.
(a) of queen, (b) under side of worker, (c) upperside of same,
(d) of drone.

On the other hand, the sting, which is curved in the queen and used only to fight other queens, is straight in the worker and accompanied by a much better developed poison sac, which deposits venom in the wound made.

§ 29. The **sting,** which is barbed, is used for self defense and for the protection of their home. It is composed of three distinct parts, of which the sheath or awl forms one. These three parts join near the edges, and form a tube, which viewed sectionally, ABB, has the shape of a triangle, the angles being rounded off (Fig. 17, 18).

The other two parts or lancets BB constitute the sting proper and in the sectional view are semi-circular, the upper edges being thicker than the lower ones, and squared to each other, one of the edges having a projection extending along the under or inner portion of it, thereby forming a rabbet along which the opposite part freely moves. The under or inner portion of these parts tapers down to extreme thinness, while near the termination of the edge there runs a minute groove which corresponds with a ridge T T in the sheath of awl H, and along which the parts move freely. Each of these parts properly tapers down to a fine point. In the cut, the right hand lancet is removed from the other parts to show its adjustment B, in sectional view.

Near the point begin the barbs, which in some stings may number as many as ten, (Fig. 17,), extending along the sting nearly one half of its length and are well-defined. Each of the lancets, when the sting is in action, has an independent motion, so they are thrust out alternately and when working their way into a wound, the valves E E by their action force out poison which is received from the poison sack C through the reservoir S. When the bee stings, it may happen that one or both of the chief parts of the sting are left in the wound when the sheath is withdrawn, but are rarely perceived on account of their minuteness, the person stung at the same time congratulating himself that the sting has been extracted.

§ 30. The worker may live as long as six months or more in the winter, when she is not flying about, but in summer her life is very short, averaging less than forty days. She literally wears herself out. For that reason, a queenless colony, in which the number of bees is no longer replenished by daily hatchings soon dies. A colony which has failed to raise a queen after swarming or whose queen has been lost in her wedding flight will be entirely depopulated by fall. A workerbee never lives to see her anniversary. Those reared in the fall, having little out-door work to perform, will live till spring. None of them die of old age but the majority work themselves to death and many die from accident.

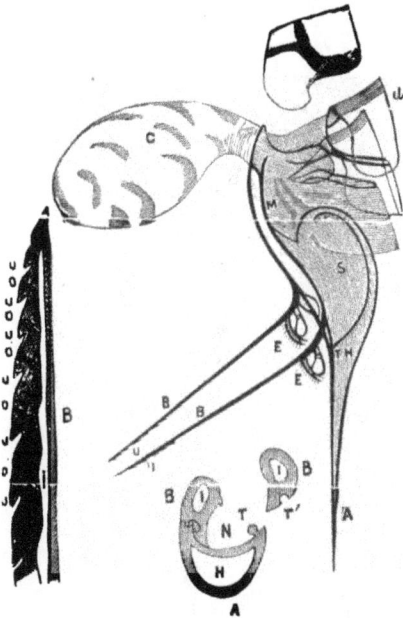

Fig. 17—Sectional View of Sting (C) poison sack, (B B) lancets of sting, (A) awl or sheath, (U) barbs, (H) hollow in awl, (I) hollow in lancets, (S) reservoir connecting with poison sacks, (T) ridges in awl.

Brood

§ 31. There are three stages to the development of the insect, whether queen, drone, or worker, before it becomes a perfect insect. These stages are; egg, larva or grub, pupa or chrysalis.

§ 32. The egg is laid by the queen in the bottom of the cell; in three days it hatches into a small, white worm, called "larva," which, being fed by the bees, increases rapidly in size; when this larva nearly fills the cell, it is closed up by the bees.

§ 33. The royal jelly (7), or white pap, which is fed to the queen larva during the entire time of its existence, and to the worker and drone larvæ during the first three or four days, is a whitish semi-transparent fluid, said by Cheshire to be produced by the large sali-

Fig. 18—Magnified Sting.

vary glands (21) of the worker bees. On the other hand, Cowan and others claim that it is a chyle product of the stomach of the worker, regurgitated by them at will.

During the latter part of their life as larvæ, both the workers and the drones are fed with a coarser food, which appears to be a mixture of pollen (55) and honey (49).

§ 34. The student will readily learn to recognize the three dif-

Fig. 19.—Eggs in the Cells (greatly magnified.)

ferent kinds of brood. The queencell (42) hangs from the edges of the comb, looking, when empty, like an acorn cup, and when sealed like a peanut. The worker-brood when sealed differs from sealed honey in its uniformity and in the leathery appearance of its capping. The drone-brood differs from it in the more rounded shape of its cappings, which (to quote one of our contemporaries, Mr. Pellett) look like bullets (Queencells, Fig. 5 and 65).

Fig. 20—Larvæ in the Cells (greatly magnified.)

§ 35. The worker develops from the egg in 21 days. The drone hatches in twenty- four days, and if the weather is propitious he will "fly" in a few days after. The queen matures in fifteen days.

§ 36. In ordinary circumstances the workerbee does not leave the hive until about eight days after hatching. During that time she is not idle. She serves as a nurse to the other young bees. At about the eighth day, she takes her first flight, and a number of them usually take flight together, in the warmth of the afternoon. They fly about the entrance, in happy and humming crowds, making peaceable circles to learn the location and appearance of their home so as to return to it without error (69).

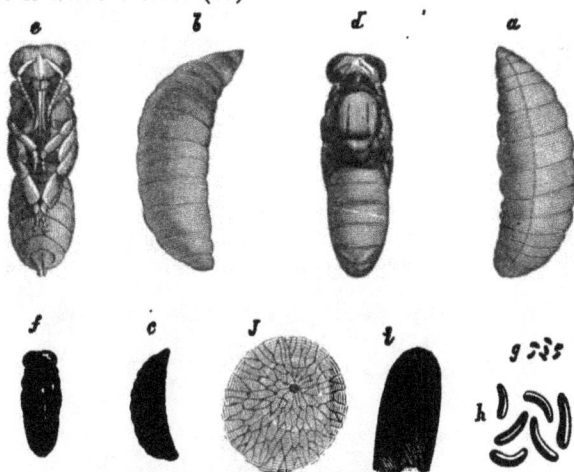

Brood in all Stages.
(a, b) magnified larva, (c) natural size, (d, e) magnified nymph, (f) natural size, (g) eggs, (h) same magnified, (i, j) micropyle or opening through which egg is fertilized.

It is of some importance for the beginner to learn to recognize the first flight of young bees, as the bustle of their pleasant sally somewhat resembles the actions of robber-bees, with which it must not be confounded. We will speak of this later (84) at the chapter upon "robbing".

§ 37. Duration of development of the brood, from the egg to the perfect insect:

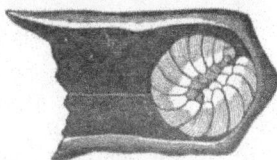

Fig. 22.—The Larva in the Cell (magnified.)

	Queen.	Worker.	Drone.
In the egg .... Days	3	3	3
Growth of larva .... Days	5	6	6½
Spinning of cocoon ...... Days	1	2	1½
Period of rest .... Days	2	2	3
Change in chrysalis or pupa Days	1	1	1
Change to winged insect . Days	3	7	9
Average duration of changes..	15	21	24

The above figures represent the time required in summer weather. In cool or cold temperatures the time is somewhat lengthened.

Wax and Comb

§ 38. The combs, hanging downwards perpendicularly from the ceiling or upper part of the hive, are built by the workers, of beeswax, produced by eating honey, much as animals produce fat, and are thus quite expensive, both in labor and material. Before the time of

Fig. 23—A Comb of Worker-brood.

Huber it was generally supposed that wax was made from pollen; but Huber fully demonstrated that bees could construct comb from honey, or sugar syrup without the aid of pollen. However, a few years previously, Duchet, a French writer, had already suggested that beeswax was produced from honey.

§ 39. The production of wax, from which the comb is made is one of the most remarkable phenomena of the organization of the honeybee. The segments or rings of the abdomen of the bee overlap each other, six in number. At the underside of four of these rings are pairs of five-sided, clear, transparent surfaces (Fig. 24), on which the plates or small scales of wax are formed, by a peculiar process of digestion. Each worker-bee is therefore able to produce eight (Fig. 25) small scales of wax. The queens and drones are not supplied with these organs. When a swarm of bees is about to leave its old home

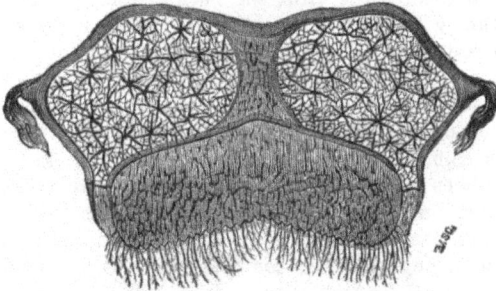

Fig. 24—Wax Producing Organ.

and seek another, each bee fills itself with honey. After entering their new home, the gorged bees suspend themselves in festoons, hanging from the top of the hive. They hang motionless for about 24 hours. During this time the honey has been digested and converted into a peculiar animal fatty product, which collects itself in scales or laminæ beneath the abdominal rings (Fig. 25). This is the wax.

Langstroth remarks as follows on this subject:

"It is an interesting fact, which seems hitherto to have escaped notice, that honey-gathering and comb-building go simultaneously; so that when one stops, the other ceases also. As soon as the honey-harvest begins to fail, so that consumption is in advance of production, the bees cease to build new comb, even although large portions of their hives are unfilled. When honey no longer abounds

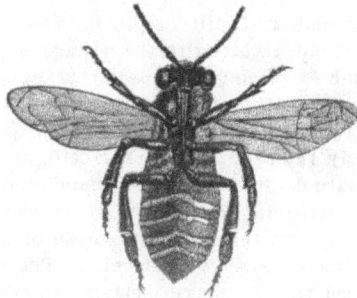

Fig. 25—Under Surface of Worker, Showing Wax in Segments (magnified.)

in the fields, it is wisely ordered that they should not consume, in comb-building, the treasures which may be needed for winter use. What safer rule could have been given them?"

The explanation of this fact by natural causes is very easy, and demonstrates the fitness of Nature to all cases. The production of wax is involuntary in the bee, whenever it is compelled to remain a long time with a stomach full of honey. Its production, which is imperceptibly small in ordinary circumstances, increases rapidly as soon as conditions demand it. As long as there are plenty of empty cells in the hive to receive the crop, the bees are not compelled to retain honey constantly in their stomachs, and there is only enough wax produced to repair or elongate the cells and seal them. But as soon as the want of room compels many of the bees to remain filled with honey for twenty-four hours or more, a sufficient amount of scales of wax is produced to build combs to store the surplus honey.

§ 40. As colonies usually swarm only during a good honey flow, many of the bees composing the swarm often have scales of wax already produced under the abdominal rings. This explains the great speed and apparent cheapness of production of comb in such circumstances. The honey required for this production was supplied previously.

§ 41. The cells, hexagonal in shape, are built on both sides of a midrib or base, and their adjustment, made in the most economical way that Nature could devise, is such that the base of each cell composed of three lozenges, makes the one-third of the base of three opposite cells (Fig. 84). The greatest economy of space and labor, combined with the greatest possible strength of construction, is evidenced in this work. The cells in which the worker-bees are reared measure about five to the inch, or a trifle over twenty-seven to the square inch. The cells in which drones are reared, and of which about ten per cent are built in the brood apartment, measure four to the inch, or about eighteen to the square inch. The total for both sides of the comb is, of course, double that number.

The above figures are not of regular exactness. There are slight differences in the sizes of cells. European experimenters have variously reported the worker-cells as numbering from 736 to 854 to the square decimeter. The former number was held preferable in supplying artificial foundation, as larger bees could be secured, in larger cells. But the difficulty arose of an occasional production of drones in these larger worker-cells. The experiments of Mr. Langstroth, reduced to the metric system, showed 838 worker-cells to the square decimeter as the standard number.

A number of cells are built, which are called intermediate cells, when changing from worker to drone-comb (Fig. 28). These intermediate cells, or cells of accommodation, are of irregular shape, and of sizes varying between the other two, according to requirements.

§ 42. Besides the cells already enumerated large cells, hanging downward and shaped like an acorn or a peanut, are found here and there, especially at the edges of the combs (Fig. 5). These are queen-cells (34). In them the queens are reared for swarming or to replace

Fig. 26—A Comb Covered With Bees.

the old queen when she becomes unfertile. The worker and drone cells are used not only for brood-rearing, but also for storing honey. Pollen is almost invariably stored in worker-cells.

§ 43. Queencells seem to be always built from particles of the comb on which they hang (probably because they are built afterwards), and are of the same color, even if newly built on old combs.

§ 44. The thickness of worker combs is about an inch, with a space for the passage of the bees of about seven-sixteenths of an inch, down to five-sixteenths. As these distances may be slightly increased without troubling the bees, we usually place the combs in our hives one and a half inches apart from center to center. It increases the case of manipulation (92)

§ 45. The cells are not placed horizontally, but incline slightly downwards from front to rear.

§ 46. It is estimated that from seven to fifteen pounds of honey are required to be consumed by the bees to produce a pound of comb. The quantity undoubtedly varies greatly according to the conditions in which the bees find themselves when the comb is built. The greatest amount is secured during a strong honey-flow, in a summer temperature. Excessive heat is objectionable only, in this connection, when sufficient to render the wax too soft and cause a break-down. "Blood heat" is undoubtedly the most satisfactory.

Fig. 27—Drone and Worker-brood Sealed.

The great cost of wax to the bees has caused apiarists to devise methods whereby the beeswax produced from combs that have been melted may be returned to the bees in the shape of comb foundation, forming the base of the comb, which will be mentioned in a separate chapter (132).

§ 47. At first when the combs are built, they are generally transparently white, but with age and use for brood-rearing they become dark and opaque. The thin cocoons lining the cells help to make them so; such are, however, just as valuable for breeding purposes for a long time, or until the size is materially diminished. They are also valuable for storing honey, where the extractor is used. During a harvest of honey and pollen of deep yellow or amber shade, the comb promptly assumes that color, though white when first secreted.

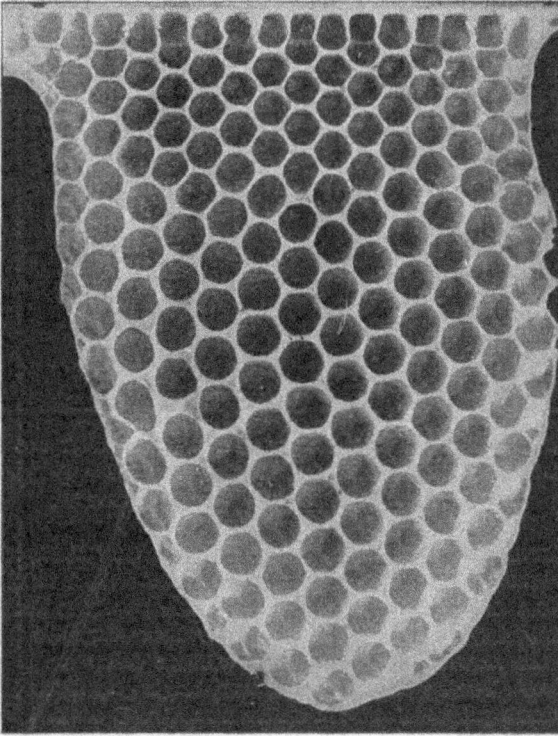

Fig. 28—A New Comb, Worker and Drone-cells.

Propolis

§ 48. Propolis is a resinous substance gathered by the bees from the buds or limbs of some trees. It is brittle in cold weather, but so sticky in the summer that the bees apply it immediately to the purposes for which they procure it. It is used to stop the cracks in the wood of their home, sometimes to reduce too large an entrance and often to strengthen the combs at their junction with the walls. They also use it to cover up obstacles which they are unable to remove, such as partly rotten wood in the hollow trees which they in-habit, or to cover up and in some manner embalm the bodies of large insects or the bones of mice which have found their way into the hive and have been stung to death by them.

Fig. 29—A Brood-Comb with Honey at the Top, Sealed and Unsealed Worker-brood Below and Queencells Along the Edges—the usual Appearance in the Breeding Season.

It is conveyed to the hive, in the pollen baskets, (27, Fig 15, 16) in the same manner as pollen, already mentioned. Such is the intelligent wisdom of those little insects that they rarely gather propolis when honey may be had. They evidently know that the honey will waste in the flowers if not removed without delay or is it mainly gormandize? During dull, dry summers, large quantities of propolis are sometimes gathered and the entire inner walls of the hive thickly coated with it (82).

Food of Bees

§ 49. HONEY is a vegetable sweet, produced mainly in the nect-ariferous tissues of a large number of blossoms. It is not made by by the bees, but only gathered, although during the transfer from the blossoms to the hive, in the stomach of the bee, it undergoes a slight change, by the action of the bee's saliva upon it. So it is customary to call it "nectar" before it is gathered, and "honey" when it has been placed in the cells.

Its color, taste, and smell differ according to the blossoms from which it is harvested. The color ranges from water-white to very dark brown. The taste and smell almost invariably indicate the blossoms from which it has been gathered.

§ 50. It contains more or less water, (163) according to the season in which it is produced, the atmospheric conditions, the amount of moisture in the soil and the kind of plants from which it is harvested. It is said that the fuschia produces exceedingly thick honey. Also, the nectar gathered from the heather, in dry, sandy plains, is often so thick that it is extracted from the comb with difficulty. On the other hand, white clover honey is often so watery, when first gathered, that it drops from the combs, like water, when they are handled by the apiarist. It sometimes contains from 75 to 90 per cent of water. But the bees soon ripen it by the warmth of their hive and the ventilation which they give it by the fanning of their wings. During a heavy harvest, rows of bees may often be seen, facing the hive and extending from the edge of the alighting board into the hive and all through its combs, their wings moving with such rapidity that they are invisible. This serves not only to evaporate the honey but to give pure air to the inside and keep down the temperature of their busy home.

§ 51. The quantity of nectar produced by the blossoms differs as well as the quality. Its flavor also differs as does its color. The whitest honey is gathered from white clover, alfalfa, basswood blossoms, etc., the darkest from late flowers such as the spanish needles or bidens, goldenrod, buckwheat, boneset, etc.

§ 52. In addition to the nectar of the blossoms, the bees also sometimes gather and bring to their hives a sweet substance, called honeydew, which is mainly produced by aphides or plant lice, although in some instances it is an exudation from the leaves of some trees or plants.

Much speculation was indulged in, some years ago, concerning the manner in which the honeydew could be dropped upon the leaves of trees even in the uppermost branches. It was ascertained that not only do the wingless lice produce a sweet substance which often attracts ants, but the winged lice also produce and eject this sweet substance, while on the wing, so that honeydew is often found on the upper surface of leaves which have no visible trace of lice above them. This sweet is of very low quality and not properly acceptable as honey. But since the beekeeper cannot avoid the gathering of it by his bees, during some seasons, the sale of it must be tolerated, for manufacturing purposes.

§ 53. The quantity of honey harvested by a colony of bees, in a single day, may vary from a few ounces to twenty pounds or more. But a large portion of this gain is promptly evaporated, as the honey thickens. Twenty-five per cent of the weight of fresh nectar usually disappears during the first twenty-four hours.

§ 54. When the honey is sufficiently ripened, the bees seal or **cap** their cells with a thin covering made principally of wax.

§ 55. **POLLEN** is the fertilizing dust of flowers. We have said, in the Preface, that the visits of insects are needed for the fertilization of the flowers. On the other hand, the pollen which the bee gathers and helps to spread upon the pistil, to fructify the seed, is needed in the economy of the hive. It serves to make the coarse pap or food given to the worker and drone larvæ (33) in the latter part of their existence in that stage of insect life. It is also consumed by the bees, during their active life. Cowan and Cheshire, following the German scientists, have interestingly described the "stomach-mouth" which appears to sort out or sift the pollen grains contained in the honey brought to the hive in the honey sack (23); for much of the honey harvested by them contains small pellets of pollen floating in it.

During the winter, however, when the bees are inactive, the need of pollen is not felt. At that time honey only is suitable and such as contains the least quantity of pollen is the healthiest, for it leaves very little residue in their intestines during the long confinement. Honeydew, fruit juices and dark honey loaded with pollen grains, are bad winter food (186).

§ 56. In early spring, as pollen is required for larval food, if there is a scarcity of it it may be replaced by flour or meal, supplied to the bees, in open boxes in well-sheltered places (Fig. 30), on warm days.

Fig. 30—Flour in Lieu of Pollen.

The flour must be packed down with the hands into a lump, so the bees will not smother in it. A shallow box or a dish are equally good. To attract the bees to the spot a little old comb may be used. Early supplies of pollen, however, are sure to be had wherever the soft maple, hazel, willow, etc., are found.

§ 57. Pollen is stored in the worker-cells of the brood combs, closely surrounding the brood. It is rarely found in the supers (90) or storing apartment. Its presence there lessens the value of comb honey for table use.

§ 58. Let it be remembered by the horticulturist that the bees are needed in the fertilization of most of his fruits. Not only do they cause the pollen to be spread upon the blossom which they visit, but in their flights they carry it from one blossom to another and thus bring about cross fertilization and greater fructification.

Water

§ 59. The use of water in preparing the food given to the larvæ is necessary at all breeding seasons when fresh watery nectar is not

Fig. 31.—Water With Cork Chips.

at hand. In early spring, bees are often noticed around well pumps and watering troughs, unless some stream close at hand gives them a constant supply. But for the use of the adult insect, water is not needed. It is well however to see that it may be had within a short distance of the apiary during cool spring days when the life of the workers may be endangered by too long trips. A bucket or pail (Fig. 31) with floating cork chips is the handiest watering trough, as the bees are in no danger of drowning.

Establishing an Apiary

§ 60. The intelligent lovers of nature admire the habits of the bees, their skill in extracting the nectar of the flowers, their preference for the best honey, for they never seek for cheap sweets, sugars or syrups, if nectar is to be had; their eager ejection from their home of dead bodies of their own race or of other insects, which if they cannot drag away, they will carefully cover up and entomb in propolis;

Fig. 32—Bees on a City Lot.

their love of cleanliness and quiet; their singularly clean management and handling of so adhesive a liquid as honey, from which they issue forth as if they had had nothing to do with it; the careful making of their combs, remodeling to suit themselves even the pretty comb foundation furnished to them; their orderly policy, their love of home; their apparent indifference to anything regarding themselves which is not for the common good, throwing themselves into danger and fighting for their hive at the loss even of life.

Beekeeping as a Science

To succeed in any calling, we must first gain a reasonable amount of knowledge of the science upon which are founded the rules of that

Fig. 33—Miss Emma M. Wilson, a Noted Illinois Apiarist and Writer on Bees.

art. Beekeeping is a science, having for its object the attainment of a correct knowledge of all that pertains to the habits and instincts of these wonderful insects. Therefore, to make the pursuit both pleasant and profitable we must possess the requisite knowledge of the laws that govern these industrious creatures. The lacking of these things will account for the many failures of those whose enthusiasm is not

Fig. 34—Model Swiss House-apiary.

Who Should Keep Bees?

§ 61. The care of an apiary is well adapted to furnish recreation to men of sedentary professions, lawyers, ministers, doctors, and teachers especially, who are often at leisure during the summer months when the bees require the greatest amount of attention, and who may thus add quite a little to their income. Ladies may keep bees, and often succeed better than men, because they pay more attention to details. "The bee-business is a business of details." It has been stated that the handling of heavy hives or supers full of honey is too hard for women, but it is easy to secure occasional help to do the heavy lifting required only when the honey crop is good.

Patience, persistence before discouragements, neatness and foresight are the requirements of an apiarist. You must also learn to handle bees without fear, if you expect to enjoy the work. This is not difficult, and directions will be found in another chapter (74). Very little capital is required, for the business must be learned on a small scale.

Suitable Location

§ 62. Unless you expect to keep bees on a large scale, almost any location is suitable for an apiary.

Fig. 35—Bees on the Roof of the Coal Shed.

Mr. Newman's apiary was located in Chicago, close to one of the main thoroughfares and street-car lines, and the result in both increase of colonies and honey was exceedingly satisfactory. Mr. Muth and Mr. Weber, of Cincinnati, had their apiaries on the roof of their store—and were successful with them.

Other apiarists without number have succeeded with bees, when owning only a very limited area of land in the suburbs of large cities. House-apiaries are resorted to in congested locations. In Europe and especially in Switzerland the house-apiary (Fig. 34) is often considered indispensable. Though it is less convenient for handling the hives, making artificial increase and for many other operations, it has the advantage of better protecting the hives, preventing robbing and requiring less space.

It is a fact that even around large cities, many good honey-producing plants are found, in vacant lots or about deserted streets. Sweet clover abounds around many a large city, and two-thirds of the honey harvested about Chicago is of this source.

A noted beekeeper of Missouri for years kept an apiary of from 60 to 100 colonies within the suburbs of the city of St. Louis. His

Fig. 36—Sheltered on the North.

bees often flew over the city and across the Mississippi river, about two miles, to work upon the bloom across this stream and harvested large crops.

If, however, you wish to make bee-culture a specialty it is best to make choice of a location where fruit and flowers abound, where white clover is found in the pastures, and fall blossoms in the fields.

Don't go where there are already many other bee-keepers, for several reasons:

1st. If you should have Italians, you don't want to have your queens fertilized by impure drones.

2nd. The pasturage may not be sufficient to support more bees.

3d. Older beekeepers may think you are "treading on their toes." and it may lead to unpleasant feelings, and a disastrous competition. A territory of three or four miles all alone is quite a luxury, if you intend keeping bees for profit.

A timber range is desirable, for a large portion of their honey and pollen they may gather from timber and shrubs. Many good localities are found near rivers or streamlets, where abound linden, sumac, maple, willow, cottonwood, and other trees, shrubs and vines that yield honey and pollen.

Fig. 37—Apiary Well Sheltered from Winter's Wind.

§ 63. The bees should be near the house, or where they can be heard when they swarm. They should be so located that the north winds would not strike them, where they can have a warm, calm place to alight.

A hedge, high board-fence (Fig. 36) or building on the north and west are a protection against the strong winds which destroy very many laboring bees in the spring (183), when one bee is worth as much as a dozen in the latter part of summer, as they are then much needed to care for the brood and keep it warm.

If, in April, the day has been rather warm and the evening cool and windy, hundreds of bees may be found on the ground in front of the hive, perhaps loaded with pollen, but exhausted from the flight and chilled with cold. As they approach the hive they relax their exertions, and a light whiff of wind dashes them to the ground, from which they are perhaps unable to arise, and before the sun could warm them up, the next morning, they will be dead.

§ 64. If you have no shade for your hives, it would be best to plant fruit-trees among them. These not only supply pollen and honey in blooming time, but acceptable shade in hot summer days (86).

§ 65. Use sand, gravel or cinders under and around the hives, to prevent the springing up of grass to the annoyance of the bees. Sawdust is also used but is dangerous as it may get afire from the bee smoker.

Which Way Should Hives Face?

§ 66. There seems to be no facing superior to the one that al-

Fig. 38—Hive Shaded with Roof—Good for Sun or Rain.

lows the sun's rays to shine directly into the entrance of a hive during the winter. There is not a difference of any consequence between a south, southeast or southwest aspect, and selection may be made to suit the apiarist's notion. Next to this, we should say, face to the east; if this is impossible, the west— and when no other available, submit to a north frontage.

Early in the spring is the best time to begin.

How Many Hives to Begin With?

§ 67. Purchase strong colonies of bees from some reliable breeder or dealer, and in order to get experience, increase from two or three colonies—not more.

Moving Bees

§ 68. Spring is the best time to transport your bees or move them from one locality to another. Select a cool day in March or April. They may be moved at any time, even in hot weather, but the danger of smothering the bees or having the combs break down is much greater. In April, when the combs contain the least amount of honey, and before the hives have become populous, bees may be moved by

Fig. 39—Shade with Roofs and Orchard Trees.

simply nailing the cap or cover and bottom-board to the main body, and closing up the entrance with a slat also nailed fast. Sufficient air will be supplied through the cracks of the entrance to keep them from smothering for several hours, and perhaps several days if they are shaded from the sun's rays. In hot weather it is necessary to remove the bottom-board and replace it with a frame fitted with wire-cloth protected with slats so as not to be in danger of having a hole punched in it in handling.

In very hot summer days, or in southern latitudes very strong colonies must be made still safer in transportation by using the wire-cloth at both bottom and top, so the bees may have a current of air.

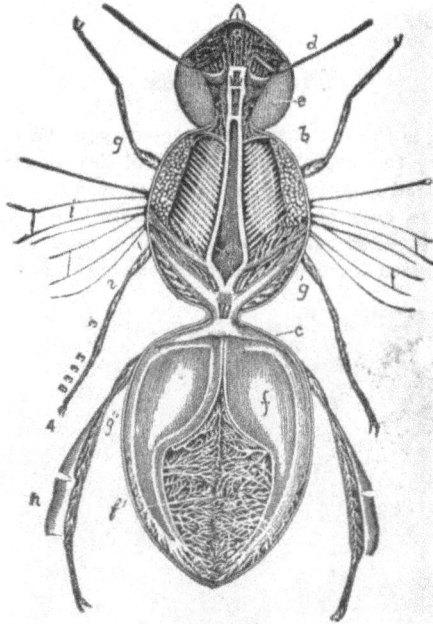

-Fig. 40—Respiratory Organs of the Honeybee.   d, antennæ; e, eyes; g, legs; f, air-trachea.

They should be kept carefully out of the sun while being transported.

§ 69. When bees are transported a very short distance, less than two miles, there is a possibility of the old bees returning to their old location. To avoid this, drum them and frighten them well before releasing them, and place a shade-board or some sort of obstruction in

front of the entrance, so they may be compelled to notice the change
of location at the first issue from home. In this way there will be
but little loss. The manner in which bees mark their location is as
follows:

They do not leave the hive in a straight line, but go only a few
inches, then turn their heads towards the hive and oscillate back and
forth in front of it; then moving further back, still hovering in front

of the hive, with their heads towards the entrance, ocasionally ad-
vancing towards it, as if to note more particularly the place of entrance and its immediate surroundings, they then increase the dis-

Fig. 41—Box-hive Apiary.

tance, taking a survey of buildings, trees, fences, or other noticeable
objects near by, after which they return to the hive, and start in a
direct line from it. On returning, they come directly to the hive and
enter; the surrounding objects and the color of the hive are all noted
by them (36).

So carefully is the location of their entrance learned by them
that if the hive is moved but a foot, they will be likely to miss their
landing at the first trip.

What Kind of Bees to Get

§ 70. Some prefer to purchase black bees in box-hives, and then transfer them to movable-frame hives in order to get experience. In that case, they should be populous colonies with the comb yellow or brown, not too dark, and as straight as possible, for greater ease in transferring. But the best satisfaction for a beginner may be obtained by purchasing strong Italian colonies with movable frames (90), in the spring. The combs must be hanging straight in the frames, or the work of transferring them would be greater than from a box-hive. Such colonies will often pay for themselves, if well cared for, in a couple seasons, thus proving cheapest in the end. One such colony is usually worth two of the former.

Fig. 42—Concrete Stands.

§ 71. To examine a box-hive, incline it to one side, or turn the bottom up, looking between the combs. By using a smoker, the bees may be driven back, and one may discover if it has capped brood, larvæ and plenty of bees. It should have such, to be considered in good condition.

Colonies which have numerous bees flying in and out in a warm day of spring, and where many bees are seen returning with pollen on their legs, may be safely considered good.

Buying "Swarms of Bees"

§ 72. A first swarm is always to be preferred, and if possible from a colony which gave a swarm the previous year, for then the queen will be in her second year—vigorous and at her best. A small, second swarm should be passed by, in purchasing. The old queen always accompanies the first swarm.

Shade for Hives.

§ 73. Whether natural shade is lacking or not, it is well to have the hives protected from the rays of the sun with a movable roof (Fig. 39, 44.). In honey-producing apiaries, where appearances are not indispensable, good roofs are made from dry-goods boxes. They are quite sufficient to protect the hives from the rays of the sun and also efficiently shelter the bees from rain. When elegant appearance is desired, roofs with two slopes, cottage-fashion, are made.

Fig. 43—Bees Filled With Honey Are Not Inclined to Sting.

Handling Bees

§ 74. The safe handling of bees is not so difficult a performance as many uninitiated imagine.

To begin with, it is important to know what are the requirements to render bees tractable. Bees that are filled with honey are **never** inclined to sting (Fig. 43). So the harvest worker returning **loaded** from the field is harmless, unless actually hurt and you may **catch** such a bee on the wing as you might do with a fly and emprison it in your hand without danger, provided you do not press it in **your** fingers.

Fig. 44—Handling Bees.

In order to render the bees of a colony tractable it is only necessary to frighten them so as to compel them to fill themselves with honey. The bees that are most to be feared are the guards which usually station themselves at the entrance to protect the hive against intrusion. When these are alarmed and compelled to retire within the hive, it is easy to overcome the possible anger of a colony.

The Bee Smoker.

§ 75. Smoke is harmless and is the best thing to alarm and quiet bees. With a good smoker (Fig. 46), blow a little smoke in at the entrance before opening the hive. Give them a little more as you un cover the frames; if very cross repeat the dose, until they yield obedience; then they may be handled with safety. Handle them gently and without fear, avoiding all quick motions; such usually incite them to anger. When honey is being stored rapidly, Italians may be handled without smoke; when there is a scarcity it is not safe to do so.

Very cross colonies of black or hybrid bees may be most completely tamed by blowing smoke in at the entrance, then closing it for a few minutes, tapping the hive meanwhile to alarm them. Swarms that have left the hive to seek another abode are rarely cross because all or nearly all the bees composing them are well supplied with honey.

Veil and Gloves.

§ 76. To those who are commencing, and until familiarity causes the loss of fear, a pair of gauntlet gloves and a veil are necessary but after fear has been overcome, a good veil will be sufficient. Such may be placed over a hat, the bottom of it coming down under the coat or vest, and when thus adjusted it is a complete protection for the neck and face (Fig. 45).

A pair of gauntlet rubber gloves is best for those who need such protection, while unaccustomed to manipulating bees, but they are cumbersome at best The advanced apiarist prefers to have the free use of his hands at all times. Bees when often handled become accustomed to the practice, and when this is gently done, they will scarcely notice the disturbance.

Fig. 45.—Bee-Veil.

§ 77. On being stung, brush the bee and the sting away as promptly as possible, because by so doing you may prevent most of the poison from emptying itself into the wound. The muscles which surround the sting have a spasmodic action, which causes pressure of the poison-sack and a deeper driving of the sting into the flesh. It is therefore a great mistake to hesitate in promptly brushing off the bee and sting with a sweeping motion which forces the sting and poison-sack away from the spot.

Remedies for Beestings.

§ 78. Dozens of different ingredients have been recommended as a cure for beestings. Most of them are useless. The poison of the bee is extremely volatile and diffuses itself in the blood promptly. Usually the practicing apiarist becomes inoculated by the poison and in a few days suffers but little from its effects. But some persons are very sensitive and suffer greatly. There may be fever and increase of temperature, in which case cold water applied to the swollen **parts** will prove beneficial. On the other hand there may be faintness **and** depression, when the administering of a small amount of stimulant will be necessary. Very nervous and excitable persons may be bene-fitted by taking a dose of bromide of potassium. The action in each case will be upon the nervous system, which must be either strength-ened or quieted. Nature is thus helped to work the poison off, which, is, after all, very infinitesimal in quantity.

§ 79. The poison of the bee has been used satisfactorily as a remedy for rheumatism, in constantly increasing doses, beginning at first with a single sting, doubling the number every other day. So the novice will understand that the bee poison rarely proves danger-ous.

How to be Safe

§ 80. While going about the apiary, do not stand in front of the hives when unnecessary, as you interfere with their flight; do not make any quick motions. If a bee threatens you by flying about your head in quick and threaten-ing circles, stoop down quiet-ly and walk slowly out of the way. Never jar a hive until you have given it a few puffs of smoke at the entrance. A great deal of the annoyance which beginners experience in managing their bees comes from their carelessly jarring neighboring hives while handling others. That is why the placing of several colonies upon the same stand is unadvisable. The least jar caused by the opening of one colony arouses the neighbors.

Fig. 46—Bee-Smoker.

Do not wear black or woolen clothes about the apiary. Wool is a fuzzy texture of animal origin, while cotton is a vegetable product resembling the stems of plants among which the bees seek their sustenance. A man is safer from stings dressed in a light-colored thin cotton or linen shirt, even with his sleeves rolled up to the elbow, than in the thickest of woolen garments.

A Hive Tool.

§ 81. Among the few inexpensive implements necessary in the management of the apiary is a "hive tool" of some sort. This implement (Fig. 47) is needed to pry the frames apart, or the different stories of the hives when opening them, since they are always more or less glued together by propolis (48). It is also handy to scrape away bur-combs that are often built by the bees, when the hive is crowded, between the combs or between the different stories. These bur-combs, when removed, should be rolled into a ball and kept until the time comes for rendering beeswax.

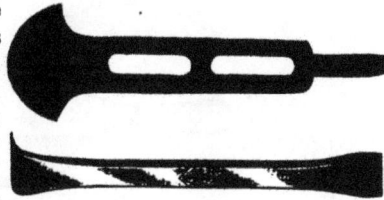

Fig. 47—Ideal Hive-Tools.

Removing Propolis from the Hands

§ 82. During warm weather the propolis (48) or bee-glue is soft and when handling frames of hives which have contained bees for a season or more, much of it will stick to the fingers, and it is rather difficult to remove.

Wood alcohol or turpentine will remove it readily, as also will slack lime kept in a small quantity in the bee house. If turpentine is used, rub the hands lightly with it until the glue is dissolved, then wipe the hands with a dry rag or paper before washing them in suds. Alcohol has the advantage of leaving no unpleasant smell behind.

Removing Bees from the Combs or Frames

§ 83. In the manipulation it is often necessary to remove the bees from the combs. The following is the "shaking off process" as practiced and recommended by Mr. G. M. Doolittle:

"Place the ends of the frame on the ends of the two middle fingers of each hand, and then, with a quick upward stroke, throw the ends of the frame against the ball, or thick part of the hand, at the base of the thumb. As the frame strikes the hand, let the hands give a sudden downward motion, which makes the shock still greater. As the frame strikes the fingers again, it is thrown back against the hand, and so on till all, or nearly all, of the bees are off. The principle is that the bee is on her guard all the while to keep from falling off thus holding in tenaciously so as not to be easily shaken off. By the sudden stopping of the upward, and a quick downward motion, the bees are thrown off their guard and dislodged from the comb. I do not remember ever having a broken comb by shaking it, as here described. If we disturb the Italians, causing them to fill themselves with honey, they can then be shaken from the combs about as easily

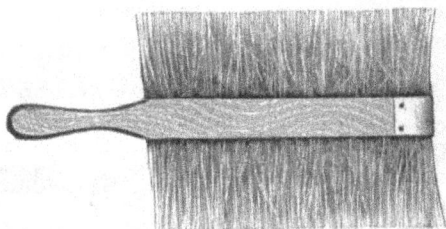

Fig. 48—Whisk-Broom for Brushing the Bees off the Combs.

as black bees. But even if we cannot afford time to wait till they are filled with honey, four-fifths of them can be shaken off."

The remainder may be brushed off without trouble by using a whisk-broom (Fig. 48). Vegetable fibre irritates bees less than animal matter. Therefore a fibre broom is preferable to a quill or a goose-wing.

Robber Bees

§ 84. In the chapter upon "brood" (36) we have spoken of the flight of the young bees, as somewhat resembling the actions of robber bees. It is now time to mention this danger.

During times of scarcity, if you see bees flying about the corners of the hives or around the entrances with a quick, hurried motion, acting as if always afraid of being pounced upon by others, you may be certain that they are robber bees. These are not a special kind of bees, for in times of scarcity all bees may become robbers if any inducement is given to this trait of their nature which prompts them to hunt for food wherever it may be had and to take it away from weaker bees or colonies wherever they find them. The sneaking guilty manners of the robber always point it out. However, should one colony succeed in overpowering another, they soon consider the prize their own and lose the guilty look, but their hurried actions and sleek, shiny appearance always cause them to be easily detected.

Remedy for Robbing

§ 85. If all the colonies are kept strong there is no danger of robbing. It is only the weak ones that are robbed. Working with bees at unseasonable times, leaving honey exposed in the apiary, etc., induces robbing. Colonies of black bees and queenless hives or nuclei (107) are usually the sufferers. Contracting the entrance, so that but a single bee can pass, is usually a cure for robbing. In times of scarcity of honey, the apiarist should be careful not to keep a hive open long, or robbing may be the result. All strong colonies maintain sentinels at the entrance in times of scarcity. Those of that colony

Fig. 49—Natural Swarming—Cutting Down the Swarm.

are allowed to pass, but strangers are "arrested on the spot." If a colony is unable to defend itself, close up the entrance and remove it to the cellar, or some other convenient place, for a few days, and when it is returned to the old stand, contract the entrance to allow only one bee to pass at a time.

Another very good method, when robbing has just begun, is to throw a bunch of loose grass over the entrance. The hive-guards station themselves in that grass and arrest the robbers that are bold enough to try to enter. But this method is unavailable after the bees have once given up to robbers. In that case, if the colony is worth saving, the only salvation is to find the robbing colony by sprinkling a few of the robbers with flour and exchanging one hive for the other, placing the robbed colony on the stand of the robber, and vice-versa. The behavior of the bees in such case is quite ludicrous, as they find themselves fooled by the exchange. Such means should be resorted to only in extreme cases.

§ 86. The best way to avoid robbing is to leave no honey exposed in unprotected places.

We have said that it is only the weak colonies that are robbed. There are exceptions, however, and in accidental instances we have seen very populous colonies overpowered. The breaking down of combs of honey from excessive heat (64) or the injudicious exposing of their stores by the apiarist in times of scarcity may induce this.

§ 87. The sudden robbing of a colony through accident is often indicated to the experienced apiarist without even having to go to the apiary, as the bees become so excited by the mutual information which seems to spread like wild fire, that some plunder is to be had, that they fly about the apiary and about the house yard with great diligence and this flight does not at all resemble the quiet hum of field workers going back and forth to the honey harvest.

Hives

What Hive to Use

§ 88. A good hive gives the apiarist complete control of the combs: It must give sufficient room for the breeding apartment as well as for surplus honey, and must admit of close scrutiny and easy manipulation.

The Langstroth Hive

§ 89. Though movable-frame hives were in use in Europe, in rude form, as early as 1795, they were not at all practical until our own distinguished and honored Langstroth, in 1852, presented the world with one that has, with his system of management, completely revolutionized beekeeping everywhere, making it a practical science.

With the movable-frame hive, all the combs can be taken out and replaced, or exchanged with other hives at will, without the least detriment to the bees. The combs having a surplus of honey can be emptied with the extractor (161), without injury, and returned to the hive to be refilled—thus saving labor for the bees in making new combs, and honey for their keeper.

Fig. 50—Original Langstroth Hive

The queen can be found, examined, and, when necessary, can be replaced by one more prolific, or one in some other way more desirable; and artificial colonies can be made by dividing at will, as we shall see hereafter (111). If a colony be weak, it can be strengthened by giving it a frame or two of brood from some other hive or it may be fed by supplying it with combs of honey from wealthy colonies. In fact the movable frame enables the beekeeper to perform any operation he may see fit to do and control the condition of his bees and their increase.

Since the invention of the movable-frame hive, many hives of different styles have been devised, but the principal feature of this hive has been retained in nearly every instance, to wit: a hive containing frames which are spaced from the body of the hive, about ⅜ of an inch, on ends, bottom and top. This space prevents the bees from gluing the frames to the body of the hive with propolis (48), and makes them removable at all times, provided the comb has been

L. L. LANGSTROTH
Fig. 51—"Father of American Apiculture"—(1810-1895.)

built straight in them. This straight building of comb was formerly secured by a triagular edge on the under side of the frame top-bar, from which the bees hang their combs. It is now almost invariably secured by the aid of strips or full sheets of comb foundation (135).

Hive Details

§ 90. The modern movable-frame hive is composed of the following parts; **bottom-board** or floor; **brood chamber** or body, containing a certain number of **frames; supers** or storage room, in which either frames or honey sections are used; **cover** or roof.

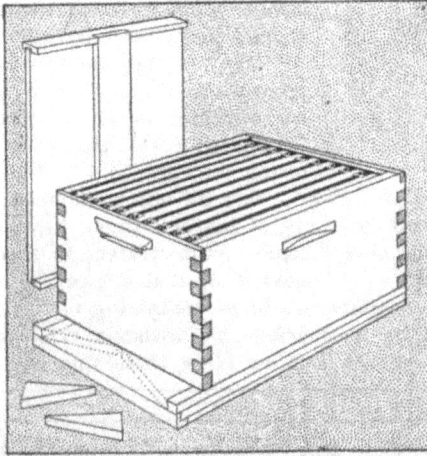

Fig. 52—"Dovetailed" or Lock-Cornered Hive

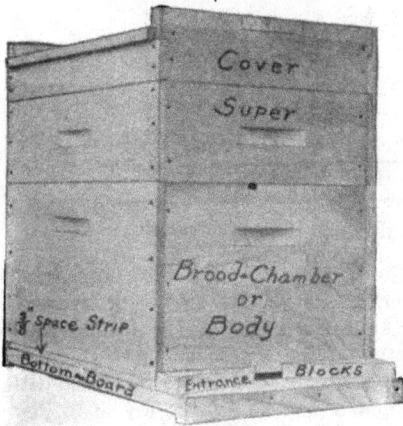

Fig. 53—Halved Corner Tri-State Hive.

§ 91. Hives are made dovetailed or lock-cornered as in Figs. 52 and 54, or halved as in Fig. 53.

The lock-corner hive, though more tightly fitting when first built, will last less time than the other, owing to its many joints exposed to the weather. Both hives are nailed from both sides, and unlikely to warp if well painted.

§ 92. The plain hanging frame is used mostly, and is the easiest handled in the brood-chamber (Fig. 55). But some people use the Hoffman self-spacing frame (Fig. 56), which is a little better liked by beginners because they cannot make the mistake of putting too many or too few frames in each hive. The combs of the bees are spaced from 1⅜ to 1½ inches from cénter to center (44) within the

Fig. 54—Hive-Body
With Plain Hanging
Frames.

brood-chamber (Fig. 53.), and closer or farther spacing will result either in too narrow and imperfect combs, or in two combs being built on the same support, making undesirable irregularities (Fig. 57).

Many frames are now made with two grooves on the underside of the top-bar, one of which is for receiving the comb foundation (141), the other a wedge to fasten it (Fig. 58).

The all-important requirement in the use of movable-frame hives is to have the combs built straight in them, and this is secured by that device.

§ 93. Most apiarists use hives containing ten brood-frames of Langstroth size, or measuring 9⅛ inches x 17⅝ outside. Some apiarists, however, use smaller hives containing only eight frames.

Fig. 55.—Plain Hanging Langstroth Frame.

The writer is much in favor of the large hives (Fig. 116), and uses a still deeper frame—11½x18½ inches— as success cannot be expected permanently unless the hives are sufficiently spacious to accommodate the most prolific queens at the time of the breeding, previous to the honey crop. Not only must the hive brood-chamber contain cells enough for all the eggs that the queen may be able to lay in 21 days (which is the period required for the worker-bees to hatch), but it must also have space enough to hold, in addition, enough honey (49) and pollen (55) for their needs.

A very important recommendation must be made here. Whatever hive you decide to use, have but one size, one style, not only in number of frames but in all parts where interchange may become necessary. You must be able to exchange, not only frames of combs containing brood or honey, from one hive to another, but supers (Fig. 53) or upper stories, sections (148), bottom boards or covers (Fig 53).

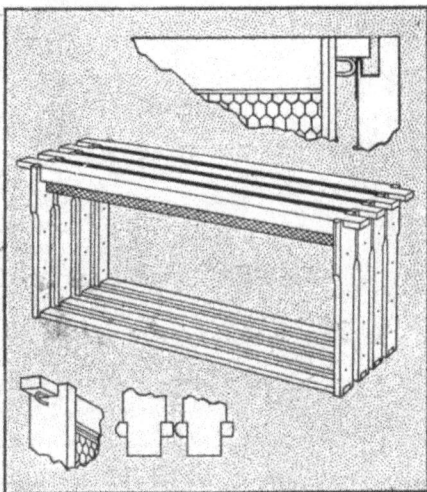

Fig. 56—Hoffman Frames.

Most of the success of an apiarist is due to his being able to exchange one part of a hive for the same part of another hive, without hesitancy and without delay. So, get well informed as to the style of hive and fixtures best suited to your location and use no other, until you see fit to change your entire system. The writer has paid dearly for the experience which prompts this advice.

Transferring Bees

§ 94. Transferring bees from box-hives is not so common as it was fifty years ago, for very few bees are now hived in boxes or "gums." The best time for this is during fruit-bloom, when the hives are not overrunning with bees, and are light in honey. A good day should be selected when the bees are at work on the bloom, as there is consequently no robbing (87).

After smoking the bees at the entrance of a box-hive, remove it some distance from the old stand, leaving an empty decoy hive or

Fig. 57—Queencells Built for Swarming on a Comb that was
Spaced too Far from its Neighbor.

box in its place, to receive the bees that return from the fields; in-
vert the hive, place an empty box or hive over it, of the same size
and shape, wrapping a sheet or cloth around where they come together, leaving no cracks large enough for a bee to escape. By
smartly tapping or drumming the hive for some time, most of the
bees, with the queen, will enter the upper box.   When they have
nearly all left the hive, place the upper box with the bees on the old
stand removing the decoy. Being alarmed and filled with honey,
they may be handled without fear.

The old hive may now be removed to a convenient room or
building, and taken to pieces, by cutting off the nails with a cold-
chisel and prying off the ends, cutting the combs when taken out as
near as possible to the size of the frames to be used.

It is well to lay a frame over the combs spread upon a large
board, so as to cut them to fit exactly (Fig. 59).

Previous to this work, the apiarist should have prepared a num-

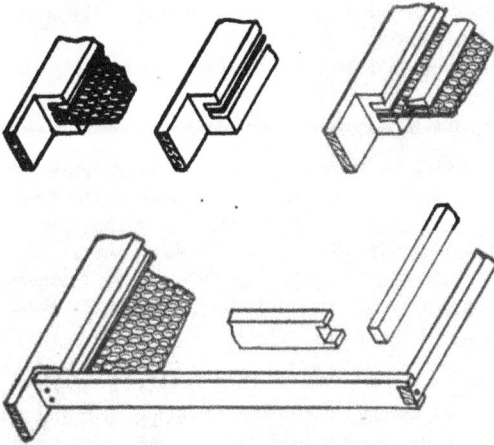

Fig. 58—Insertion of Foundation in Frame.

ber of wires of proper length to be fastened by a bent end into the top and bottom bars of the frames. A number of these wires (Fig. 60), are driven on one side of a number of frames, and the frames, one at a time, laid wire side down on the board. Then the comb is fitted in and other wires nailed on top. The frame with the brood has then the appearance of the one in Fig. 61. Twine may be used, but it stretches, and does not hold the comb as stiff as wire. A previous edition of this book advised the use of sticks fastened with wire. The method we give here is much the best.

No drone-comb or drone-brood should be transferred. The bees will always have too much. Chickens will eat the drone-brood if it is given them. The worker-brood, of course, should be transferred so as to occupy

Fig. 59—Cutting Combs To Fit Frames.

a central position in the frames as is natural in the hive. The honey is placed in the rear of the combs.

Carry the new hive to the old stand, and empty the bees out of

Fig. 60—Bent Wire to Fasten Comb in Transferring. The ends are driven in the upper and lower edge of the Frame.

the box on a sheet, in front of the hive. See that the queen, as well as all the bees enter it. To prevent robbing the entrance should be contracted; and in a week, when the bees have fastened the combs, the transferring wires should be removed. Always work slowly with the bees, and avoid jarring.

Short Method of Transferring.

§ 95. For the benefit of the apiarist who does not wish to do as difficult a job as the one just described, we will give an easier method, though less economical, of transferring the bees from box hives, or from movable frame hives in which the combs have been built crooked and are therefore immovable.

Instead of performing the operation at the blooming of fruit trees, it is delayed till early swarming time. A new hive with frames full of foundation or of comb already built is placed on the stand of the colony to be transferred. The bees are then driven as above directed and shaken in front of this new hive, taking good care that the queen is among them. This virtually makes an artificial swarm. Enough bees are returned to the old hive to enable them to take care of the brood and it is placed in close proximity of the new hive usually a little behind it. At the end of ten days destroy the queen-cells (6). After twenty-one days, the brood being hatched out of the combs (37), the bees are shaken in front of the new hive and the old hive and its combs en-

Fig. 61—Appearance of the Transferred Comb.

tirely taken away to be handled at leisure. These combs contain only honey, which may be extracted (161), as the apiarist may see fit. If advisable, the bees of this old colony may be used to strengthen any small colony in the apiary. Being mainly young bees they will remain where given (36).

A Very Large Swarm.

Swarming and Queen-Rearing.

Natural Swarms.

§ 96. Swarming is the natural way of increase for bees. It usually takes place in May or June, in the North.

For some days before swarms issue the bees may be seen clustering at the entrance of their hive, though some come out where there are little or no indications of a swarm. When honey is abundant, and bees plenty, look for them to come forth at almost any time, from the hours of 10 in the morning to 3 in the afternoon, for first swarms; for second and third swarms, from 7 in the morning until 4 in the afternoon.

By examining the colony it can be ascertained whether they are about to swarm. If queencells (6) are seen with eggs or larvæ nearly ready to be sealed over, a swarm may be expected within one or two days after the first cell is sealed over, or as soon after as the weather will permit.

After whirling a few minutes in the air, the mass of the bees will cluster on the branch of some convenient tree or bush—generally one that is shaded from the sun's rays.

Fig. 63—A May Swarm

They should be hived as soon as the cluster is formed, else they

may leave for the woods; or, if another colony should cast a swarm while the first are clustered, they would probably unite.

Should the queen fail to join the bees, by reason of having one of her wings clipped, or for any other cause, the swarm will return to the hive as soon as they make that discovery. As the bees are gorged with honey, they may be handled without fear of stings (Fig. 64).

Fig. 64—Bees Peaceable When Swarming.

§ 97. "Afterswarms" being unprofitable, all but one of the queen-cells should be destroyed, or cut out—this will usually prevent any more swarms issuing. Within eight days, the first queen will hatch and will take possession of the hive.

§ 98. The queen has very little to say as to swarming preparations. If a second swarm is desired by the workers they will prevent the first queen (6) from destroying the other queen-cells, which are sometimes very numerous, even on small pieces of comb. (Fig. 65). She would be sure to do this if not hindered in her desires. If thus restrained, she will show her irritation by "piping," and this piping is answered by the other queens which are kept prisoners in their cells. The second swarm then issues within two or three days.

After the departure of this swarm, and the emerging of the second queen, if her "piping" is also answered by a third queen, a third swarm may also issue.

If the desire to swarm is satisfied after the departure of the first swarm, all the queencells will be destroyed by the first young queen that emerges. The worker-bees often help her in this task, which she performs with great alacrity and energy.

How to Hive a Swarm.

§ 99. If the cluster be low it is easily performed. Have a hive in readiness, slightly raised from its bottom-board in front. A sheet is spread in front of it. The limb on which the swarm is hanging may be cut off and the swarm carried to the hive and shaken down on the sheet. By watching for the queen she will be readily noticed and directed towards the hive. She will enter it eagerly for she loves darkness. The bees will crawl into the hive, and finding the queen, be satisfied to remain. When the bees are in, place the hive where it is to remain; a shaded position is the best. If comb foundation (Fig. 85), be placed in the frames, it will be of very great advantage in comb-building.

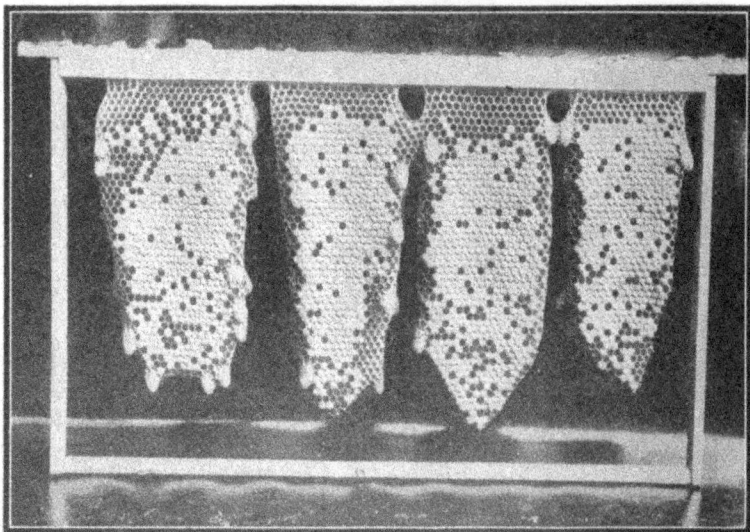

Fig. 65—Dr. Miller's Method of Producing Queencells.

If the bees have clustered on a branch or twig, which is too valuable to be cut down, a basket, box or swarm sack (Fig. 68), will be quite essential, into which to shake or brush the bees. If on a wall or fence or on the trunk of a tree, brush them into the basket and proceed to hive as before described.

Sometimes the swarm is placed where it is even impossible to brush the bees into a box, or perhaps the number which may be gathered together at one time is so insignificant that they take wing at once to return to the cluster. Then, if a comb of brood may be had from another hive, or even if a frame full of dry comb can be used, by placing it over the swarm, the bees will soon recognize it as of possible use to them and will readily cluster upon it, or they may be gently smoked so as to direct them towards it. But care must be taken to not frighten them away.

Fig. 66—Catching the Queen as She Issues from the Mother Colony.

If the colony is noticed in the act of swarming, the queen may be caught as she emerges from the hive and put into a cage (126, Fig. 66, also 131) and allowed to run in with the swarm, (Fig. 67).

If two swarms have taken wing at the same time, and cluster together, they may be evenly divided by placing two hives on the ground and directing the bees equally to both, especially if the queens are found and caged and placed at the entrance of the hives. Bees may be scooped or shaken without much trouble when the swarm is gathered and they are easily directed to one hive or another if properly handled.

A frame of brood placed in the new hive will be of much advantage to the bees. It will prevent the swarm from leaving the hive, and should the queen be lost it will give them means of rearing another. By filling the other frames with comb foundation (135), they will soon be in good condition and perfectly at home in their new quarters.

Fig. 67—Letting the Queen Run in the New Hive.

Sometimes a swarm will go to the woods without clustering—but this is rarely the case.

The beating of tin pans is, of course, of no avail; throwing a stream of water from a fountain pump is often done to bring down an absconding swarm, and cause them to alight and cluster.

Afterswarms.

§ 100. Secondary or afterswarms are usually not desirable. If they issue, they should be gathered in the usual way. At the end of 24 to 48 hours they may be returned to the parent colony by shaking

Fig. 68—
Swarm-Sack.

them in front of the entrance. It usually does away with any further swarming for that season. The primary swarm might be returned in the same way, if no increase is desired, but its remaining in the hive is far from being a certainty, as the hive is still too populous for comfort.

Prevention of Natural Swarming.

§ 101. As many beekeepers do not wish to indefinitely increase the numbers of their colonies and as, on the other hand, some beekeepers have other occupations which prevent them from watching

Fig. 69—Combs Built in Open Air.

the apiary to gather the swarms, it is often desirable to prevent the issue of swarms, as much as practicable.

Entire prevention of natural swarming is a goal which cannot be reached. But in numerous instances it may be greatly reduced. The following requirements are to be fulfilled.

1. The colony must be placed in a spot where the bees will suffer as little as possible from the heat. Hives exposed to the direct rays of a hot June sun are uncomfortable. The bees often will be seen clustering on the outside, unable to remain within, even with the help of numerous fanning workers who may be seen busy cling-

Fig. 70—Suffering from the Heat.

ing to the front board and vibrating their wings to force a current of air through the crowded hive.. The roofs mentioned at the chapter on "Shade for Hives" (73) are very useful in this connection.

2. There must be ample room for ventilation. If the ordinary entrance is not sufficient, the hive should be lifted from its bottom-board, and blocked up an inch or two, especially in front.

3. There must be room for the depositing of the crop.

We have shown at the chapter on honey that as much as 20 pounds of honey may be harvested by a colony in a single day. If our bees are crowded for room, they will at once make preparations for swarming.

4. A very limited number of drones should be permitted to be reared. Drones are cumbersome and noisy and they are in the way of the workers (12). If we destroy the drone-comb, leaving but a very small amount and **replace it with worker-comb** (139), we will have more workers, less drones and less incentive for swarming.

5. The queen must be young and provided with ample space to lay. An old queen is often the cause of swarming because the bees, noticing her decrease of fertility, make preparations to replace her by rearing queencells (6). If the season is favorable, she will leave with a swarm as soon as the queencells are sufficiently advanced.

6. The spacing of the frames 1½ inches from center to center is more favorable to swarm prevention than the 1⅜ inch spacing (92), because the bees have more room to cluster and better ventilation between their combs when the wider spacing is used. Most of the modern hives made have the narrower spacing.

Queen-Traps.

§ 102. Although the fulfilling of all the above-named conditions will not entirely prevent swarming, it will have a great tendency to decrease it. Other means may be taken, such as destroying the queencells by making regular weekly visits of the combs, or using a queen-trap so that the queen cannot leave the hive (Fig 71). A queen-excluder or bee entrance guard (Fig. 72) may be used which will entirely prevent her egress from the hive. But these contrivances are in the way of the bees. However enticing the use of them may look to the inexperienced they are usually discarded by practical apiarists.

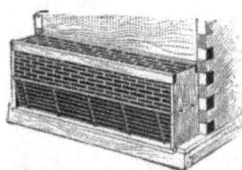

Fig. 71—Drone-and-Queentrap.

§ 103. Some experienced apiarists claim success in the prevention of swarming by lifting all or most of the brood into an upper story divided from the lower story by an excluder, giving the queen empty combs or comb foundation in the lower story and removing this upper story later. Some place this story of brood on top of the supers. Those methods are efficient only in-as-much as they give the queen extra room for laying eggs in the lower story and supply the bees with

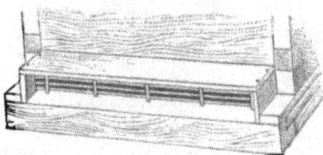

Fig. 72—Bee Entrance Guard.

more space for honey.   They are practical especially with small
hives.   But the placing of brood combs over a super filled with honey
sections will result in more or less darkened cappings in the sections, probably because the bees carry bits of combs from the upper
brood chamber down to the super, which very rarely happens when
the brood is below.

Artificial Increase.

§ 104. As the apiarist does not always wish to wait upon the
pleasure or whim of the bees to increase his apiary, methods have
been devised for artificial increase, or by dividing colonies of bees.

Queen-Rearing.

§ 105. We have explained on another page (6) that when the
colony becomes queenless from any cause the workers at once proceed to rear another queen, provided they have eggs or larvæ less
than three days old.   In order to make artificial increase it is first
necessary to build queencells, if none of our colonies are preparing
to swarm.   It is only sufficient to remove the queen of a colony in
order to compel them to build queencells and as shown in Fig. 65,
they often build a large number.   At the end of nine days, we are
ready to use these cells.

Fig. 73—Plain Division-Board or Dummy.

§ 106. It is necessary to say here that if we wish to secure the
best results from our bees, we should breed our queens from the
colonies that are the best producers.   As a rule they are selected
by their results in the previous season.   A pure race is best, as
it is more likely to keep true to its traits than hybrids.   This subject will be considered further in the chapter on Improvement in
Honey Bees (122).

Nuclei.

§ 107. The next thing is to make nuclei. These are made by taking two or more frames, as may be desired (at least one of which should contain brood), with adhering bees, and shaking into the hive the bees from one or more additional frames, so that there may be enough young bees to remain after the old bees have returned to their former hives, to keep the temperature sufficiently high to hatch out the brood, as well as to care for the emerging queen. In making up nuclei be sure not to take away the queen with any of the frames.

G. M. DOOLITTLE
Fig. 74—Author of "Scientific Queen-Rearing," a Well-Known Beekeeper and Writer.

It is better to use the regular frames for nucleus hives, and either use the ordinary hives with a division-board or dummy (Fig. 73), to contract the brood-chamber, and economize the heat, or make small hives just to suit the number of frames used.

These nuclei having been made on the ninth day after the starting of queen-cells, and in such number that there may be one queen-cell for each and one for the mother colony, we cut out these queen-cells on the tenth day, and insert one in each nucleus. The queens will hatch shortly after this, and we have saved the bees some labor and time.

§ 108. To cut a queencell out, commence on each side of the base of the cell, not nearer than half an inch, and cut upwards a wedge-shaped piece, the cell in center, being careful not to squeeze or even handle the base of the cell. A similar wedge-shaped piece must be cut out of the frame of comb that it is desired to put the cell into. Then carefully place the cell into the hole thus made, fitting it securely in position; place the frame in the hive and close it up.

§ 109. As the virgin queen emerges from the nucleus to meet a drone, sometimes the bees will accompany her if they have no unsealed brood. To prevent this, two or three days after the queens are hatched, insert a frame containing eggs and young larvæ in each nucleus. If the queen should be lost on her bridal tour, the materials will be on hand for the bees to rear another, if it is unnoticed by the apiarist.

§ 110. In two or three days the queen will be hatched, and a week or ten days later will become fertilized, and be laying; this may be readily discovered upon examination. Now the apiarist is ready for the formation of new colonies, without the inconvenience of natural swarming, by

Dividing the Colonies.

§ 111. Bees swarm because it is their natural manner of increase. By dividing them we secure the increase without swarming, and save time in watching and hiving natural swarms. This, however, must not be overdone. The beginner sometimes imagines that by dividing he can make almost any number of colonies from each one, forgetting that **strong** colonies are the only ones that accomplish anything. Dividing should never be done unless the colony be very populous, and can well spare the bees and combs. To **double** the number of colonies each season is not good, unless increase is desired at the expense of honey.

Some divide their strong colonies equally, or nearly so, carefully looking for the queen, putting her into the new hive, placing bees and brood in the center, filling up with frames containing comb foundation (135), removing the hive with the queen to a new location; then giving the queenless colony one of the young queens reared in a nucleus as above described or a young queen purchased from some reliable queen breeder.

§ 112. Ordinarily we prefer to make the increase by enlarging our nuclei. Take one of the nucleus hives before described (which should be of the same pattern and size as those to be divided), and remove the division-board. Then take a frame containing brood and adhering bees from each colony, placing them in the nucleus hive until it is full. Be sure not to take the queen away from any colony. The bees that will hatch out in a few days will make that nucleus a populous colony. Put a frame filled with comb foundation (135), into each hive from which the frame of brood was taken, and in a few days they will have this all worked out into beautiful comb; and, in all probability filled with eggs.

The new colony having a young and fertile queen, and plenty of bees, will soon rival the old one in the vigor of its work. Increase being secured in this way, none of the colonies are disturbed, and the bees everywhere "pursue the even tenor of their way." All being kept strong in numbers they are ready for the honey harvest, and will soon work in the section-boxes.

Dividing should be done in the middle of the day, when the bees are busy in the fields and the yield of honey is abundant.

§ 113. Another plan, practiced with success, is to take away the division-board in the nucleus hive, fill the frames with comb foundation, and exchange places with a populous colony, caging the queen of the nucleus for about 36 hours, or until her acquaintance has been made by the strange bees that come pouring into it from the fields—for bees will always return to the exact spot occupied by their home (69).

There are other and more elaborate methods of rearing queens and making artificial increase. Queen-rearing has become a specialty, and the apiarist who wishes to go into this business should read the special works on queen-rearing, especially Doolittle's "Scientific Queen-Rearing" and "Pellett's Practical Queen Rearing."

§ 114. The apiarist who has but a few colonies and does not wish to go to the trouble of rearing queens, may divide his strong colonies by shaking part of the bees and the queen into a new hive supplied with comb foundation, and placing this on the old stand, removing the old colony to a new place. This must not be done unless there are thousands of young bees hatching daily, for the old hive thus loses all its working force. A better plan yet is to make the increase of one colony out of two others by placing the queenless hive on the stand of a third colony, and putting this in a new spot. By this method you take the working bees from one colony and the brood of another to make one division. The greatest objection to allowing a colony to rear queens, which has been deprived of its working force, is that it may suffer from cool nights, and good queens can only be reared in strong colonies, at least up to the time of hatching.

Bees Recognizing One Another

§ 115. The apiarist should bear in mind that, although bees readily recognize their own colony members, most probably by the sense of smell, the bees of different hives may be mixed without any danger of fighting during a good honey crop. Bees like human beings, are evidently peaceably disposed towards those who come to them with gifts. Such manipulations as here advised would not be practicable without great precautions in times of scarcity.

§ 116. A colony which has been made queenless should always be watched from the 20th to the 25th day, as its queen should begin to lay about that time. It is a good plan to give such a colony a comb containing eggs and larvæ, about the fifteenth day, in case the queen should get lost, otherwise the bees would have no means of rearing another, and the colony might perish. Likewise a colony which has sent forth one or more swarms, should be examined later to make sure that the young queen has not been lost on her wedding flight. Many a loss of colonies which has been ascribed to the moths (214), was caused by queenlessness. The colony, having no longer any hatching bees, slowly becomes reduced in strength until the moths invade it and easily overpower it.

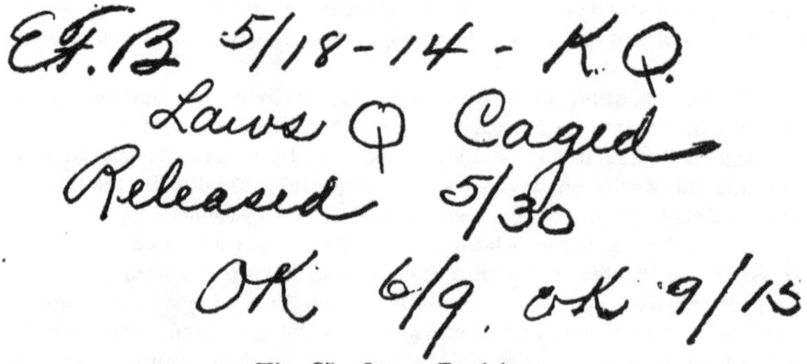

E.F.B. 5/18-14 - K.Q.
Laws Q Caged
Released 5/30
OK 6/9, OK 9/15

Fig. 75—Queen-Register.

§ 117. To remember dates everyone has not the faculty and yet all the operations of queen-rearing require that it should be done. For instance, the time when a choice colony was made queenless, to have queencells started—the time these cells are given to the nuclei—

the time of hatching—when the queens commence ᴏ lay, etc. To save time and trouble in remembering these and other dates, a small slate (Fig. 75), 3x4 inches, with a hole in the center of the top, should be hung on the hive by a small nail with all these dates written thereon. A printed card tacked on the inside or outside of the hive-cover is used by some to advantage, in keeping track of such dates. Most beekeepers have some method of keeping memoranda.

The Loss of the Queen.

§ 118. When the bees manifest a restless and uneasy disposition by running about the front of the hive and signaling each other, it is a sign that they have lost their queen, and they should be examined at once.

Should a colony become queenless from any cause, three weeks may be gained by having an extra queen to give it at once. Upon examination, if no brood is found where the bees are clustering, the colony is queenless. At any time during the season, from March to October this is a sure sign. Colonies that lose their queens during the winter have a forlorn appearance. The bees walk around the entrance listlessly and without eagerness; but few of them go in search of either honey or pollen.

§ 119. It is astonishing how quickly the bees discover the loss of their queen after she has been removed, even though there may be tens of thousands of workers in the hive. We believe that the odor of the queen pervades the hive in normal conditions and that, as soon as this odor is absent, the loss is noticed and communicated from one bee to another, for they evidently have most wonderful ways of communicating valuable information to each other though their vocabulary is necessarily limited.

§ 120. No time should be lost in giving a queenless colony a comb of eggs or young larvæ,or both, from which to rear a queen. Sometimes such a colony will refuse to build queencells: it may be too weak, its queen may be too old to lay, or they may have laying-workers. If it be too weak, it should be united with another colony. If its queen be old, she should be removed and the bees given a frame of brood from a prosperous colony. If it has laying-workers the most effective way to get rid of them is to break up the colony, dividing it among strong colonies having fertile queens.

Drone-Laying Workers

§ 121. Worker-bees being undeveloped females, it is not strange that now and then one may be sufficiently developed to lay eggs.

Prof. Leuckart remarks that "it results entirely from the development of egg-germs and eggs in the individual ovarian tubes—which proceeds precisely in the manner described in the case of the queen." As they are incapable of meeting the drones and becoming fully fertilized, their eggs will produce only drones.

Fig. 76 presents a view of the genitals of such a bee, compared with the ovaries of the queen and those of the sterile worker.

The drone-laying worker deposits eggs in very irregular manner, sometimes two or more in a cell.

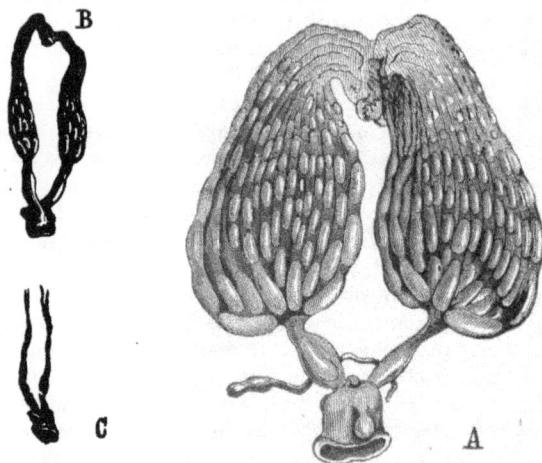

Fig. 76—Comparative Size of Ovaries.
(A.) Ovaries of Queen. (B.) Ovaries of Drone-laying Worker,
(C.) Ovaries of Worker.

If the eggs laid by drone-laying workers or unimpregnated queens are laid in worker-cells, the drones hatching from these will be diminutive in size. But they appear to be as perfect males, except in size, as the full-sized drones.

Improvement In Honeybees

Selection

§ 122. To obtain the best results we must possess the highest grade of bees that it is possible to obtain. Our object being to elevate the race, the most thorough treatment must be employed.

The queen must be prolific to be able to keep the hive full of bees, to gather the honey harvest when it comes; the bees must be industrious; they must be docile to allow the apiarist to manipulate them with ease and pleasure and they must be strong and hardy, to withstand the rapid changes in climate.

In developing the highest strain of horses, not all their offspring are equal to the best; careful selection of those coming the nearest to the ideal animal must always be made, and the closest scrutiny is necessary while making that selection. The same is true of cattle, sheep, hogs, poultry and bees. "Sports" and "variations" occur, producing inferior progeny; but all careful breeders who have an eye to the improvement of the race reject those that do not come up to the "standard of excellence," sending such animals and poultry to the shambles—so let us carefully select the best queens and drones to breed from.

Length of Tongues

The length of the tongue (Fig. 13) of the honey-bee is an important matter in her make-up, as there are some blossoms, such as red clover, which have so long a corolla that the average bee cannot reach their nectar. The Cyprian bee is said to have a longer tongue than other races, but its cross disposition renders it unfit for general domestication. The Italians have often furnished strains that harvested nectar from red clover during the second crop of that plant, and in very dry seasons, from the first crop. For that reason, we strongly recommend the Italians over any other race.

The Italian Bees

§ 123. Briefly stated, their superiority is thus demonstrated:

1. They have longer tongues and gather honey from flowers where black bees cannot.

2. They are more industrious and persevering, and with the same opportunity gather much more than black bees.

3. They work earlier and later in the day as well as in the season, often gathering stores when the blacks are idle.

4. They are better to guard their hives against robbers (84), and against the ravages of the bee-moth's larvæ (214).

5. They are more prolific in the spring.

6. Their bees and queen adhere more tenaciously to the comb.

7. They are amiable, and it is easy to manipulate them.

There are other races of bees which are desirable, and among them we will mention the Caucasian and the Carniolan, but as both of these races are of nearly the same color as the common bee, it is almost impossible to ascertain whether we have them in their purity.

8. Owing to some of the above tendencies they are more prone to overcome the dreaded disease European foulbrood (219), than the common black bee. This quality is claimed also for the two races of Carniolan and Caucasian bees, but is well proven in the Italians.

Pure Italian bees are recognized by the three yellow rings on the first three segments of the abdomen of the worker, next to the thorax or middle portion of the body. They are also singular-

Fig. 77— Italian Queen

ized by their quiet behavior on the combs when the hive is opened. If the bees are properly handled none of the Italians will rush about the combs or fall off while in the hands of the apiarist.

Their queens and drones are more irregular in color than the workers.

Italianizing an Apiary

§ 124. To do this, a tested Italian queen (Fig. 77), should be obtained from some reliable dealer or breeder, and introduced into one of the colonies of the apiary. For, as the queen is the mother of the colony, to change queens is to change the whole character of the colony in a short space of time (30).

To Introduce a Queen

§ 125. To introduce a queen successfully it will be necessary to find the queen to be superseded, and take her away. A black queen, being easily frightened, will hide or run away to some corner, therefore it is best to proceed cautiously and without jarring.

In the middle of the day, when the old bees are at work, open the hive, taking out the center frame, examine both sides, and if the queen is not seen, proceed with the adjacent frames till she is found. If not successful the first time close the hive an hour or two, till the bees become quiet, and then repeat the operation. An Italian queen is easily found, but the blacks are more troublesome. When found, either destroy her or make such other disposition of her as may be desired; cage the Italian queen and insert her in the center of the brood-chamber between two combs containing honey, which the queen may be able to reach at pleasure. This, however, is not indispensable, as the bees will usually feed her.

Fig. 78—Gentle Italians.

§ 126. The cages in which queens are mailed by queen-breeders all over the country, and especially in the South, are quite convenient to introduce queens. Those cages (Fig. 79), contain candy, and the queen is usually released by removing the stopper which gives the bees access to her by eating through the candy within the course of a day or two.

§ 127. It is well, when introducing a queen which has been received from a breeder, to allow the worker bees which accompany her to escape before placing the cage in the hive. The reason of this is that the new comers are rarely accepted by the bees of the colony, even after a day or two of confinement, except in the height of the honey harvest, while the queen is generally welcome after they have

Fig. 79—Cage for Mailing Queenbees.

become acquainted with her. The Miller introducing-cage, (Fig 80) is convenient because its flat shape permits the hanging of it between two combs, in the center of the brood. A small wire is used for this purpose. When using this cage, at the end of about 48 hours, the queen is released by removing the stopper and putting in its place a little piece of comb honey or cappings.

§ 128. During the height of the honey season, more expeditious methods may be used to introduce a queen, such as smoking the colony thoroughly after the removal of the old queen, closing it a few minutes and allowing the new queen to run in at the entrance. Queens are most easily introduced when they are freshly removed from their hive and are in egg-laying condition. So the transfer from one colony to another of a queen is attended with much less danger for her than her introduction when she has been traveling.

Fig. 80.—Miller Introducing Cage

In any case, valuable queens should be introduced by the cage method. After releasing a queen, it is well to immediately close the hive and not reopen it for a day or two, for the greatest danger to a queen comes from her being suspected when robbers (84) are flying about.

Safest Method of Introduction

§ 129. An absolutely safe method of introduction of a valuable queen consists in filling a hive or nucleus (107) with combs of hatching bees and releasing the queen from the cage upon those combs. As there are no bees except such as are in the process of immediately hatching, there is no danger whatever for the queen.

Clipping the Queen's Wing

§ 130. This is done to prevent her from leaving with a swarm. In attempting to fly she will fall to the ground in front of the hive, and the bees, missing her, will return to the hive. This must not be done until after the queen has met the drone, or she will remain un-

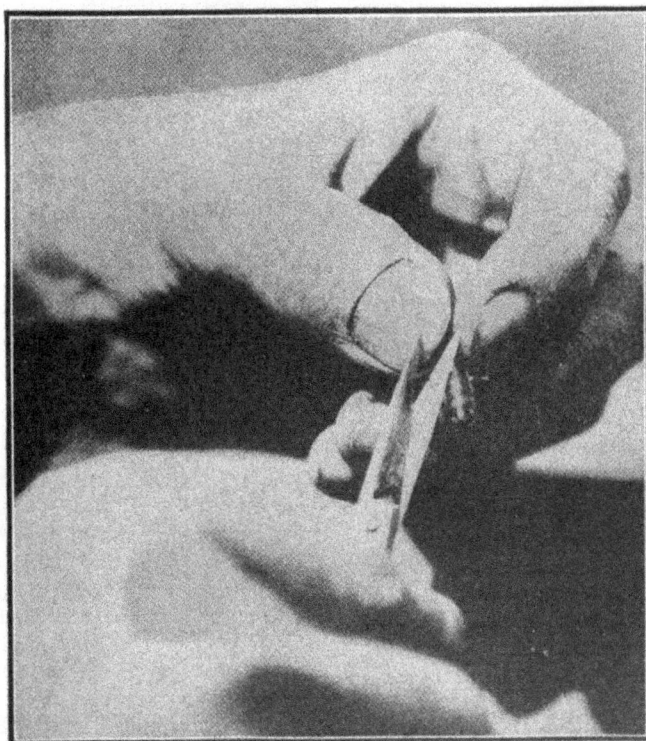

Fig. 81—Clipping the Queen's Wings.

fertile. To perform the operation, open the hive, lift the frames carefully, and avoid jars; when the queen is seen—with a pair of sharp-pointed scissors, lift one of the front wings and cut off about one-half of it. To pick her up, be sure not to take her by the abdomen. She may be held by the wings or the thorax without danger.

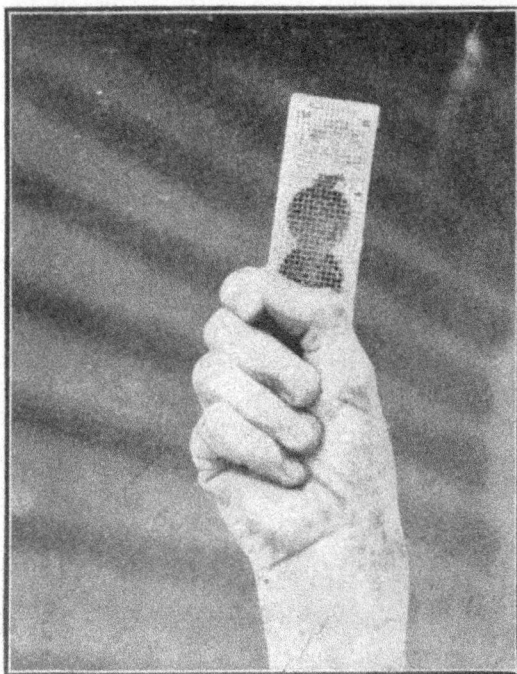

Fig. 82—Putting Queen in Cage.

Some very practical beekeepers, like Dr. C. C. Miller, the author of "Fifty Years Among the Bees", always clip the wings of their queens. It identifies them and prevents their escape with the swarm. This effectually prevents the swarm from absconding at least until a young queen is reared. But the apiary must be watched. Otherwise, the clipped queen would get lost, for she would nevertheless attempt to fly, and would be unable to return. She may usually be found, accompanied by a small cluster of bees, in the grass in front of the hive.

Dr. Miller advises, when clipping a queen, to remove both of the wings on one side of the body. This makes the queen more efficiently visible among her bees when looking for her, on account of her lopsided appearance.

Purchasing Queens

§ 131. Queens supplied by queen breeders are always expected to be fecundated and laying. **Untested** queens are such as have proven healthy layers. **Tested** queens are those whose progeny has **been examined and found of pure breed.**

When handling and caging a queen, you should let her run up to the cage, as bees or queen always travel upwards when trying to escape.

Looking for the Queen.

Comb Foundation and Its Use

§ 132. The beehive is an emblem of industry, and the perfection of its government is truly marvelous! When we view the beautiful comb so wonderfully systematic in construction, and all completed by a crowd of bees in a dark hive, we are amazed at the skill of these wonderful little architects! Think of their cells of wax, only 180th part of an inch in thickness, one ounce of which delicate work will contain a pound of honey, of sufficient strength to be transported thousands of miles without injury, with but ordinary care.

CHAS. DADANT,
Reviser of "Langstroth on the Honeybee," and an authority in both Europe and America.—(1817-1902)

The cells of the bees are found to fulfill the conditions of an intricate mathematical problem. Let it be required to find what shape a given quantity of matter must take in order to have the greatest

capacity and **strength,** occupying at the same time the **least space,** and consuming the least labor in its construction. When this problem is solved the answer is the hexagonal or six-sided cell of the honeybee, with its three four-sided figures at the base."

As the bases exactly fit into one another from opposite sides, and the insects work on both sides at the same time, in what language did they communicate the proportions to be observed, while making these bases, common to the cells on opposite sides? (Fig 84).

Cost of Comb

§ 133. We have explained that the bees consume from seven to fifteen pounds of honey to build one pound of comb (46), according to the season, the warmth of the hive, and the strength of the colony.

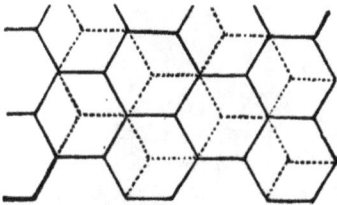

Fig. 84—Bases and Cross-Sections of Cells.

It is very evident that the amount varies much, and the comparison may be made of this wax production with the production of fat in animals. Although wax is a fatty substance, it cannot, however, be called "the fat" of the honeybee, but being produced at the expense of their nutrition, it is secreted in greater or less quantity, according to the more or less favorable circumstances in which the bees find themselves. It is therefore probable that a rule cannot be established as to the cost of wax any more than can be given as to the cost of producing fat in cattle. But starting from the amounts given above, we can safely assert that combs cost the bees, on an average, not less than ten pounds of honey for each pound of comb produced, including the time lost in elaborating it. If honey is worth fifteen cents per pound, comb therefore costs the bees the equivalent of one dollar and a half per pound. From this we may know the value of comb foundation, made from pure beeswax and returned to the bees.

Manufacture

§ 134. This comb-foundation was first invented in Germany and made from plates, by Mehring; and Mr. W. M. Hoge, in 1874, assisted Mr. Frederick Weiss, an aged German, then living in New York, to introduce it to American beekeepers. Later it was improved upon by A. I. Root, J. Vandervort and E. B. Weed.

§ 135. Comb-foundation (Fig 85) consists of sheets of beeswax, formed by dipping wooden plates into melted wax, or by other processes too complicated to be explained here, some of which consist in making endless sheets of the material which are rolled up in a manner similar to the rolling of paper for printing on cylinder presses. The sheets of beeswax are afterwards printed with the rudiments of the cells, by running them through cylinders or mills indented with the exact shape of worker-cells, and afterwards cut the proper size for frames or sections.

It would be tedious to review all the various styles of foundation presented to beekeepers since it was first introduced in America. We have had foundation with triangular-shaped cells, with flat bottomed cells, with high side walls, and with no walls at all; with linen, cotton, wood, paper, tin-foil and woven-wire for a base; we have had

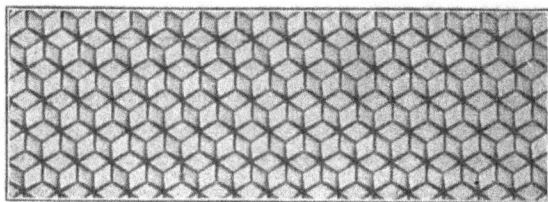

Fig. 85—Comb-Foundation.

foundation with fine wire imbedded therein, and frames of foundation with wire pressed therein.

Different Grades

§ 136. Experience has shown that the foundation which has the thinnest base is the best. The bees thin it out still more, and shape it to suit themselves. For brood-combs, sheets measuring about six square feet to the pound prove best, as the bees find in them almost

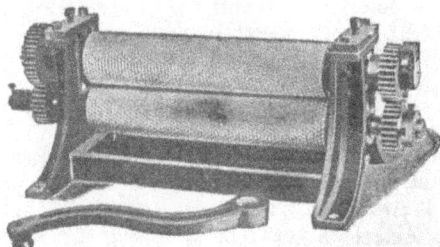

Fig. 86—Comb Foundation Mill.

enough wax to build the entire comb, especially if it is given them a little ahead of need, at a time when they have leisure to manipulate it and draw it out. The experiments of Foloppe Freres, of Champosoult, France, have proven that in drawing out the foundation into comb the bees manipulate the wax in the same way that the potter handles clay to make a vase, by "repoussage" which forces the wax towards the outer edge of the cell in a circular way. This is another

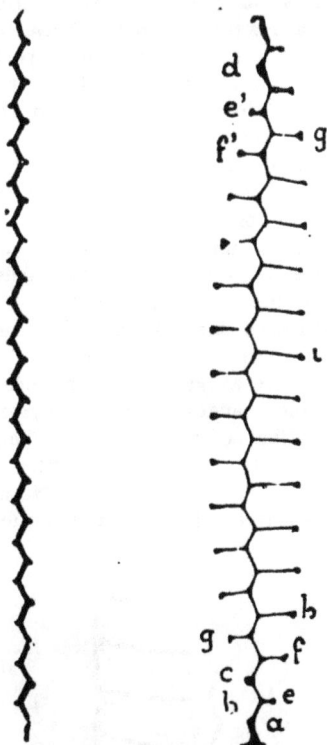

Fig. 87—Cross-Section of a sheet of Foundation—Natural size.
Fig. 88—Cross-section of a sheet of foundation in process of Construction by the bees—natural size.
(By Foloppe Freres; taken from "L'Apiculteur," of Paris.)

evidence of the bees' intelligence, for the cells are thereby made much stronger than if the drawing of the wax had been made towards the outer edge without any circular manipulation. In the same manner, if the potter had made his vase by pushing the clay out ward without the circular rotation, the vase would break much more readily.

Figs. 87 and 88 show the section of a sheet of medium comb-foundation as given to the bees and as worked by them out of the wax it contains. a, b, and d, show the manner in which the work is begun, continuing and forcing out the wax towards the edge through the different shapes assumed consecutively at c, e, f, g and h.

§ 137. For surplus honey (146) in the sections, very thin sheets of comb-foundation are supplied to the bees, and of the very best grade of light-colored beeswax, for it is important that the combs should be thin and avoid the the "fish-bone" toughness of a heavy artificial midrib. As light sheets as 13 and 14 square feet to the pound are now used in sections.

§ 138. As we have already stated, the foundation measuring about six square feet to the pound, is best for brood-combs. Experiments have proven that between five and a quarter and six feet to the pound is sufficient to supply the bees with all the material needed to build the entire comb, the cells afterwards sealed, when needed, with naturally produced wax. (Fig 89).

§ 139. The advantages derived from the use of comb-foundation are three-fold. In the first place, as we have said before, beeswax costs the bees a probable average of ten pounds of honey per pound of comb. Beeswax when rendered has an approximate value of from twenty-five to thirty-five cents per pound. The same article made into comb-foundation costs at retail from fifty to seventy-five cents per pound. Counting our honey at only twelve cents, there is almost a doubling of the investment 'by buying comb-foundation and saving the bees all this labor.

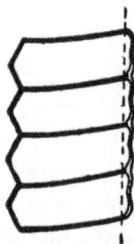

Fig. 89—One side of sheet of Foundation drawn into Comb. The Cells were made entirely from the Foundation supplied. The cappings alone have been made of natural wax supplied by the workers, as shown by the lighter lines.—(By Foloppe Freres, from "L'Apiculteur," of Paris—magnified.)

The second advantage, which is equally important, resides in giving the bees guides from which to work. Before the advent of comb-foundation, guides of different kinds were devised to compel the bees to build straight in the center of the frames or sections. But in spite of all efforts, the combs were often crooked or wavy, and irregular. With the use of comb foundation we secure combs "as straight as a board" in every frame and every section. This alone would suffice to make comb-foundation a blessing to the apiarist. No more crooked combs, no more leaking honey in handling, hence very much reduced danger of robbing.

The third advantage is almost as great as the other two. In natural conditions the bees build about ten per cent of drone-comb (Fig. 90). This is necessary in a state of nature, when colonies are far apart and the queens in their wedding flight (4) need to meet drones readily. But as only one drone is actually able to do service

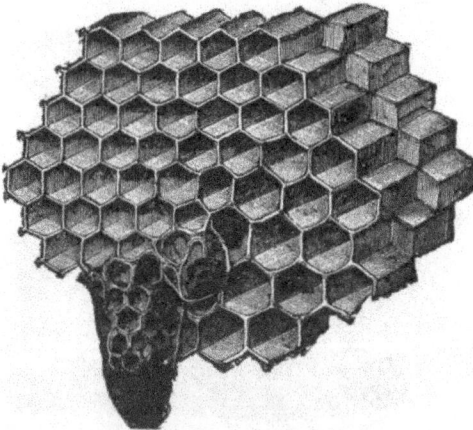

Fig. 90—Honey-Comb.
Drone-comb, Intermediate Cells, Worker-comb, Queen-cell.

for one colony, the numerous drones of one first-class colony are quite sufficient for fifty or more colonies in one apiary. Hence it is advisable to do away with the drone-comb (16, 101) as much as possible. By the use of comb foundation, made with worker-cell bases, we secure this result. Large sheets of drone-comb are dispensed with and replaced by worker-comb. We must not depend upon the bees to do this, but when we remove drone-comb we must use foundation in its place. There will always be plenty of drones reared in corners where the wax was short, or in cells that become enlarged by accident. One or two colonies with plenty of drones will be all we need, and we may select them to suit ourselves, and give them the drone-comb right in the center if we choose.

The annual saving by the prevention of rearing a horde of useless consumers through the use of worker-comb foundation is, in our opinion, sufficient to pay for the initial cost of this foundation.

The reader will readily comprehend by the above explanation why the business of comb-foundation manufacture has gained in importance. It is a product that every beekeeper needs, and he quickly realizes this.

Must Be of Pure Beeswax

§ 140. Comb-foundation must be made of absolutely pure beeswax. Its tenacity at certain temperatures; its malleability at blood

Fig. 91.—Wire Imbedder.

heat, which is the heat of the hive, make its adulteration by any other compound absolutely undesirable. The bees themselves know their product from all other compounds, and adulterations of foundation with similar products in mineral or vegetable waxes have always proven an entire failure.

Fastening Comb Foundation

§ 141. In Fig. 58. has been given the method of fastening comb foundation to the frames. For hiving swarms, full sheets are also sometimes fastened additionally with wires, and we here exhibit the method (Fig. 91). The little instrument used is called a "wire imbedder." The wire is first placed in the frames, then the foundation is inserted as in Fig. 58, and at last the imbedder forces the wire into the wax. For section-boxes (151) the Parker fastener is used, which presses the wax on the underside of the top-bar (Fig 92).

More elaborate machines are made which both fold the section and place the foundation strip in it at one operation. For the practical apiarist such contrivances are very useful.

Fig. 92—The Parker Foundation Fastener.

We show two of the most practical (Figs. 93 and 94). Many beekeepers have contrivances of their own both for folding sections and inserting the foundation. Full sheets are usually inserted. But this will be treated more fully at the chapter on comb-honey production (148).

Preserve the Wax

§ 142. The use of comb-foundation requires all the available beeswax in the country; every bit of wax and old combs should therefore be preserved. A wax-extractor (Figs. 95 and 96) will soon pay for itself. By its use all the old comb and the cappings may be saved, utilized and restored to the bees in comb-foundation to be worked out into comb, forming either the cradle of bees or the receptacle of honey.

Fig. 96 represents the best wax-extractor yet produced. It is the Hershiser wax-press. Layers of burlap in which the combs are placed with division racks between them are subjected to pressure as the wax melts in boiling water.

Rendering Combs into Beeswax

§ 143. Soft water should be used in melting wax. Iron utensils are objectionable. The iron rust colors the wax, and spoils the appearance of it in a permanent way. Tin or tinned receptacles are indispensable. But it is not absolutely necessary to have a wax-press or wax-kettle, for any ordinary wash-boiler may be used, though with more waste. Break up the combs. Soak them well in water, then heat to the boiling point, taking care not to overboil the wax, as it would both spoil it and cause some danger of its running over. Make a sort of basket or pouch out of wire cloth, and sink it into the surface of the boiling mixture. From this you may dip the wax as it comes to the surface and pour it into flaring vessels, such as crocks or tin pans. The few impurities that you may thus dip up will settle to the bottom, and the wax may again be melted to finish cleaning it.

Fig. 93—Woodman's Section Fixer.

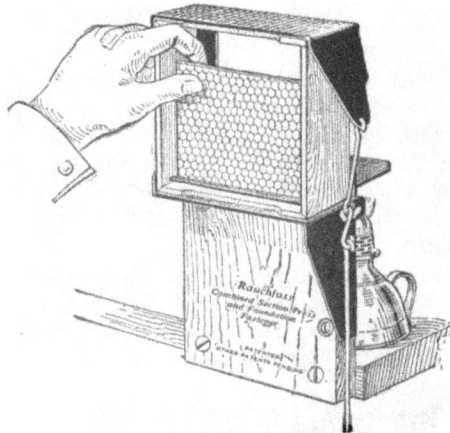

Fig. 94—The Rauchfuss Combined Section Press and Foundation Fastener.

As many people do not like to trouble themselves with the rendering of old combs into wax, the foundation makers have taken it up on a large scale, so that in many cases the apiarist can have his old combs rendered and made into comb-foundation for less than he could do it, especially as all possible waste is avoided.

Solar Wax-Extractor

§ 144. The apiarist gathers from time to time during the season a large amount of wax chips, from old broken combs or from scrapings of hives, frames and sections. If rendering with boiler and press is undesirable, he may yet provide himself with a solar wax-extractor, Fig. 97. This implement, which is insufficient and inadequate in rendering up large lots or very old combs, is quite handy to dispose of small parcels of wax. It is always at hand in the apiary. Its cost is insignificant and the broken combs thrown into it are thereby protected against the moth and may be gathered in the shape of cakes of wax at the end of the season.

Fig. 95—Wax-Extractor.

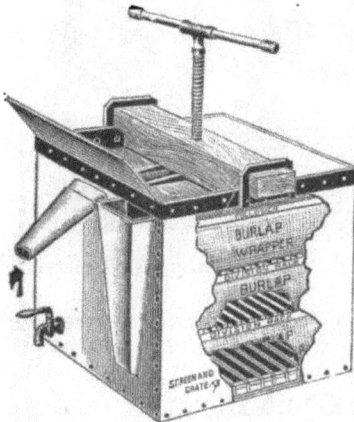

Fig. 96—Hershiser Wax-Press.

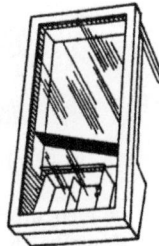

Fig. 97—Rauchfuss Solar Wax-Extractor.

Production of Choice Honey

§ 145. In no country on the face of the earth is honey produced that can excel that of North America. Nature has supplied this vast Continent with honey sources as varied as can be found anywhere in the world. And within the past few years, many improved methods and appliances have been invented for the increased production of honey. Simultaneously with these improvements, we find the consequent increased consumption.

Honey in the Comb

§ 146. Not only have we forsaken the log-gums and rude straw and box hives of our fathers, and given these busy little workers homes that are entirely under our control, but we have devised, for surplus honey, small sectional frames, so that the sealed product maybe easily taken from the hives and marketed in convenient shape, suited to the requirements of the retail purchaser.

Fig. 98—Oblong and Square Sections Contrasted.

Honey is also produced in large frames, for extracting, but we are first to describe the production for sale in the comb.

Bulk Comb Honey

§ 147. For home use and for sale in localities where small sections are deemed too expensive, honey is produced in either half-story frames or full sized brood-frames with guides or sheets of a light grade of foundation. It is a more economical way of producing honey, but not to be recommended, because of the difficulty of its sale on the large markets.

One-Pound Sections

§ 148. After numerous experiments, the so-called "one-pound section" has been accepted as the standard. Its size is 4¼ x 4¼ inches, and its width usually 1⅞ inches. The thickness of the surplus combs may be greater than that of the brood-combs given on a preceding page (92). The thickness of the brood-combs is regulated by the length of the body of the bee which is hatched in the cells, while the surplus combs may be built as thick as two or three inches as storage combs. Other sizes than the regular pound sections are used; however, uniformity is desirable rather than novelty. Three different sizes of sections are shown in Fig 98.

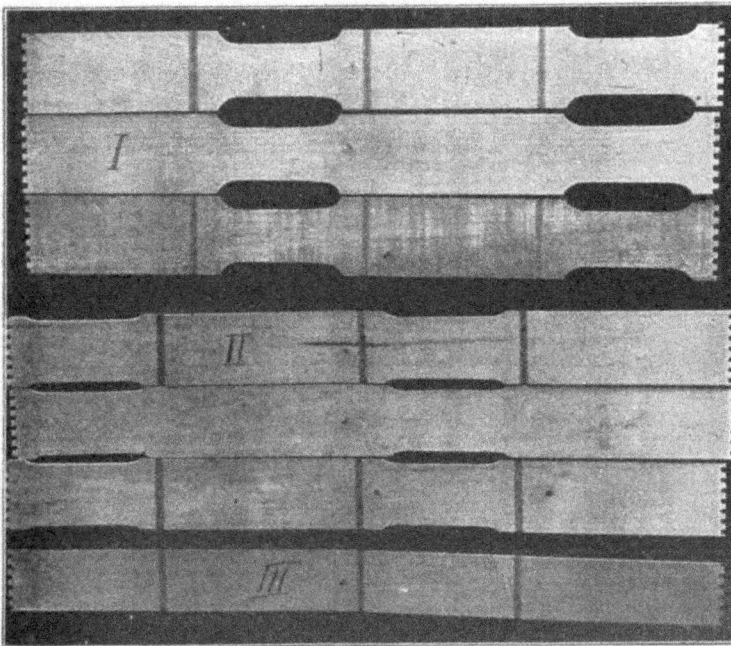

Fig. 99—One-Piece One Pound Honey-Sections.

Instead of making the section boxes in four pieces, nailed or matched together, they are made in one piece, as in Fig. 99. These can be easily bent into the shape of a box, by hand, but that can of course be done much faster by machinery, as shown at the chapter on "Comb Foundation," (Figs. 93-94).

Plain Section-Boxes and Cleated Separators

§ 149. For years the section-box has been made with one or more scallops on each edge, for the purpose of allowing the bees to enter from below, and also to pass on up to another tier of sections when supers are tiered up on the hives. But later there was introduced what is known as the "Plain" section, all the scallops being omitted, and the sections being made 1½ inches in width (Fig. 102).

In order to allow the bees to get into the sections and also pass on up to those placed over the first tier, the separators used between the rows in a super are cleated in such a way as to hold the rows of sections apart. Such separators have been called "Fences," or cleated-slat separators. Figs. 100 and 101 give an excellent idea of this separator.

The advantage of all these separators is to secure combs which do not project or bulge into each other. This allows the casing of sections from different hives within any box without danger of getting them scratched, and causing the honey to leak. Separators are not necessary when producing honey for private use only.

Fig. 100—Different styles of Fences and Separators.

Supers for Holding Section-Boxes

§ 150. There are various arrangements for holding the section-boxes in which is placed the surplus honey. Perhaps that most widely used is the section holder (Fig 105). A super used on an 8-frame hive holds six of these section-holders, and for a 10-frame hive seven of them. Each section-holder takes four sections 4¼ x 4¼ inches in size. A separator is then placed between two section-holders (Fig. 104).

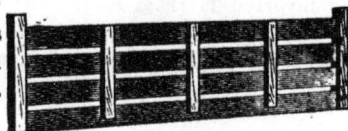

A section-holder might be called a wide frame without a top piece, simply two end-blocks nailed on a bottom slat. The section-holders are supported in the super by two strips of tin nailed crosswise under

Fig. 101—Fence or Cleated Separator.

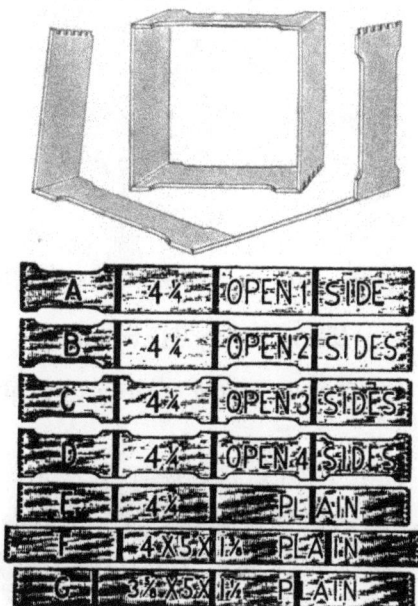

Fig. 102—Sections of Different Widths and Openings.

each end. The section-holders, with the sections and separators, are then wedged up from one side by the use of a follower-board and a wedge, thus making all snug and tight.

Miller T Super

Another very good super is that devised by Dr. C. C. Miller, one

of the most successful comb-honey producers.   It is called the **T**
super because the supports of the sections are made in the shape of
an inverted T, (Fig. 106).

§ 151. Before placing them in the hive, the sections are provided
with strips or full sheets of foundation of the lightest grade.

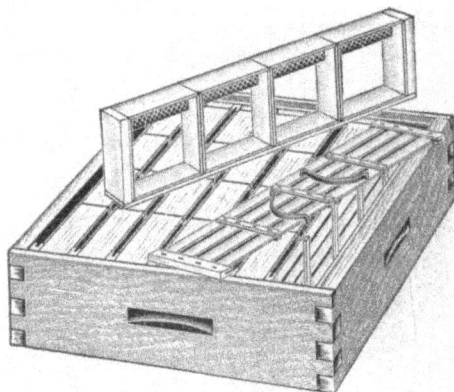

Fig. 103—Super with plain Sections, Fences and Section-Holders.

The Honey Harvest

§ 152. The time for putting the supers on the hive is when the

Fig. 104—Super of Section-Holders Filled with Section-Boxes.
Explanations—D, Solid wood separator; A, dovetailed super; E, section
boxes; F. follower-board; G. wedge for between follower-
board and super side, to make all solid.

crop begins. This may be readily recognized by the whitening of the top of the combs in the brood chamber, by the bees. In localities where the crop flow is sudden and large, it may not be advisable to wait till the combs are being whitened, for this is an evidence that the bees already gather enough honey to produce wax. Each apiarist must become acquainted with the honey resources of his region and act accordingly. Suffice it to say that every populous colony should be provided with one or more supers at the opening of the harvest. If we have any partly filled or partly built sections of the previous year, it is well to use them as bait in the center of the super,

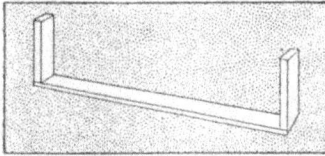

Fig. 105—Section-Holder

as the bees more readily ascend into it. But such sections, if they contain any honey at all, must have honey of the same quality as that which is expected to be gathered. It is usually best to extract such honey in the fall (165).

When Bees Swarm

§ 153. If a colony swarms, on which is a super partly filled with honey, the bees will abandon this work, owing to the depleted condition of the hive. Some apiarists have adopted the plan of removing the old colony and putting the swarm on its stand, giving it the super from the old hive at the end of two or three days. The old colony is either put upon a new spot, or exchanged with a middling strong colony which is not powerful enough to swarm, this colony being itself put in a new spot. We have already explained in the chapter on artificial swarming that during a honey crop bees from different hives may be mixed without danger of fighting.

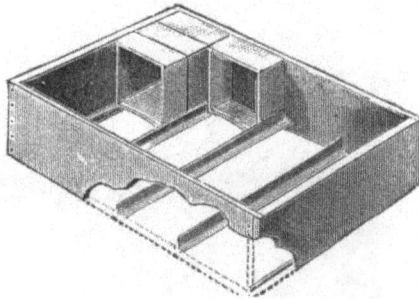

Fig. 106—T Super,

Queen Excluders

§ 154. In order to prevent the queen from going into the supers

and laying eggs there, which would soil the combs, giving them a
dark color, beekeepers have devised queen-excluders, made of zinc
with perforations, or of wood and wire that enable only the worker-
bees to go up into the supers (Fig. 109).

Fig. 107—From Comb-foundation to Sealed Honey, by G. C. Greiner.

The perforated zinc (Fig. 110) is also used for queen and drone
traps (102), to prevent the queen from going out with a swarm or in

bee-entrance guards for the same purpose. But all these contrivances are more or less in the way of the bees, and prevent easy ventilation in hot weather, and they should be avoided as much as possible.

Fig. 108—"Fancy" Comb Honey.

Large Honey Crops

§ 155. In very good honey-producing seasons and favorable locations, a colony may fill several supers (Fig. 111), sometimes as many as four or five. But it is well for the novice not to be too sanguine. As the super is filled, more may be added, either by raising the super already partly filled and adding another between it and the body of the hive, or by adding the new super on top of the other. When the new super is put on top of the other, there is less danger of the bees leaving a part of the sections unfilled or unsealed. On the other hand, the addition of a new super under the first is a great incentive to active work, and helps prevent the desire to swarm.

Have the Sections Well Sealed

§ 156. By all means care should be taken not to furnish enough room to scatter the honey in a large space, and not get the sections sealed. When producing comb-honey it is very important to get all

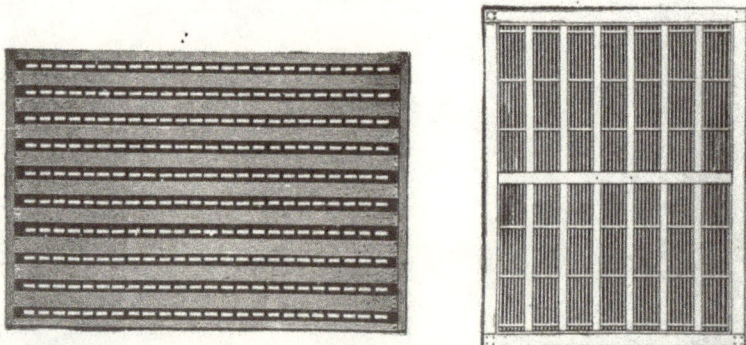

Fig. 109—Queen-Excluders to Place Between the Brood-Apartment and the Supers.

the sections fully sealed (Fig. 108). But sealed sections ought to be removed promptly (227), in order that they may not be soiled by the bees traveling over them, which discolors them and gives them a more or less stale appearance.

A few unsealed cells in a section of honey spoil its appearance and lower its grade. Extra-fancy section honey is thoroughly sealed.

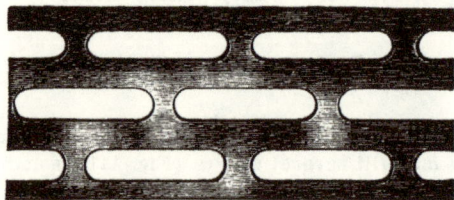

Fig. 110—Shows the Exact Size of Zinc Perforations.

§ 157. The crop of white honey should not be mixed with that of dark honey. That is another reason why it is best to have the bees fill the sections and seal them as they go. The flora of the locality must be learnt and the apiary managed accordingly.

Removing Honey from the Hive

§ 158. Before taking honey from the supers, it is necessary first to get the bees out of them. Dr. C. C. Miller does this, if the crop is still on and the bees do not rob (87), by simply raising the super

Fig. 111—Dr. Miller's Biggest Crop—4 to 7 Supers per Colony

off the hive and leaning it against the hive. The bees which are thus uncovered and exposed soon make a marching file towards the entrance. But one must watch them, as they may soon come back and begin carrying away the honey. Another method followed by him is to pile the supers taken off, covering them with a cloth in the center of which has been sewed a wire-cloth in the shape of a cone with a small hole at the top. The bees escape under this cloth to the cone and out, but are unable to find their way back. It is best not to put the piles of supers too far from the hives from which they are taken, as some of the young bees might be unable to find their way back home. However, the flight of the old bees usually indicates to them the route, and a young bee, full of honey, is welcome in any hive she may adopt, unless there is much robbing and fighting.

Piles of supers containing bees, without queens, are usually deserted by the bees shortly after they are removed if only covered with a cloth or a light sheet. The bees crawl out and away from under the sheet, but care must be taken that robber-bees do not find their way in, as they would soon carry away the honey. These operations should be performed in the shade, but during the warm part of the day, while the field-bees are at work. There is then less danger of being stung, and a less number of bees in the super. Night operations should be avoided.

The Bee-Escape

§ 159. Although some of our leading beekeepers do not use this implement, it is nevertheless one of the most practical, even though it does its work slowly (Fig. 112). This is inserted in a honey-board

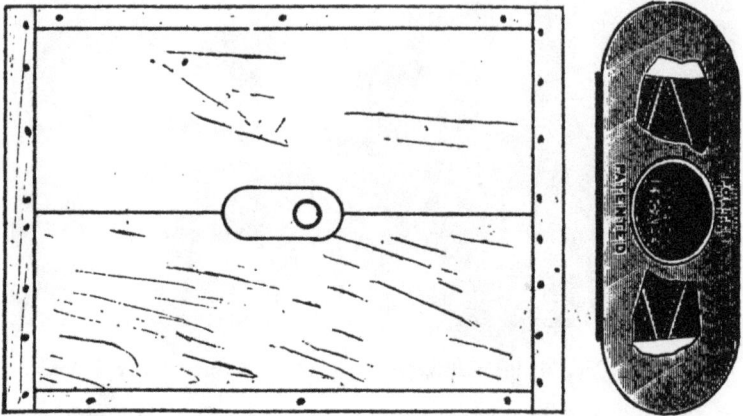

Fig. 112—Bee Escape-Board.　　　　　　Porter Bee-Escape.

between the body and the supers, and the bees which go down through it are not able to return. By lifting the supers, in the evening, and placing the escape-board under them, we are sure of finding the supers empty of bees when morning comes, with the exception of a dozen bees or so. This method is still better than the one mentioned above, but it requires two operations instead of one.

Extracted Honey

§ 160. Following closely after the increased knowledge concerning the natural history of the honeybee came improvements in beehives and modern appliances for obtaining an increased production of honey. Major Von Hruschka, a retired Austrian officer, who was then keeping bees in Italy, invented the honey-extractor and its great value is everywhere admitted by all progressive beekeepers.

The Invention of the Extractor

The following is a brief history of the discovery: One day when the Major, who was a most observing and critical beekeeper, was in his apiary, his little boy came to him. The boy had a small tin pail tied to a string, which he was swinging, boy-like, around and around in a circle, holding the end of the string in his hand. The father gave the youth a piece of comb filled with honey, putting it into the pail. The boy, after a while, began to swing the pail again as before, with the honey in it. A few moments after, he became tired of that amusement, and put the pail down to talk to his father, who took it up, and, by chance, noticed that the honey had left the comb and settled down into the pail leaving the comb perfectly clean

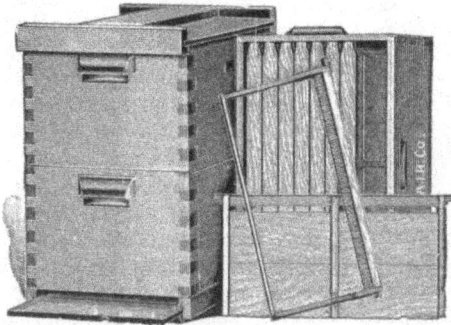

Fig. 113—Two-Story Hive for Producing Extracted Honey.

that had been on the outside of the circle when the boy was swinging it around. The major wondered at the circumstance and commenced a series of experiments which resulted in his giving to the world the first honey-extractor, which by whirling, something like nis son whirled that little tin pail, gave him the pure liquid honey, extracted by centrifugal force, leaving the honey-comb entirely free from the liquid sweet, which he gave again to the bees to fill; instead of the primitive method in use up to that time, of mashing up the combs containing the honey, pollen, and sometimes brood; to let the honey drain through the cloth in which it was placed—giving what was formerly known as "strained honey."

Major Von Hruschka's original honey extractor has been greatly improved. Now we have neat machines which do their work well and rapidly, but honey consumers generally have no idea how it is accomplished.

§ 161. Extracted honey is obtained by the combs being uncapped, placed in the basket or frame-holder of a honey-extractor (Fig. 114), and revolved; the centrifugal force throws out the pure honey from the combs, which runs down the sides of the can and is drawn off. Extracted honey is the pure liquid—minus the comb.

Supers for Extracted Honey.

§ 162. The devices employed in the production of comb-honey, sections, section-holders, separators, etc., are not suitable when we desire to produce honey to be taken out of the combs with the extractor.

For this purpose, hives are supplied with either a double story as in Fig. 113, or with half-story supers containing frames of shallow depth, as in Fig. 115.

Many people prefer the full-story supers, because they are interchangeable with the brood chamber, both in hives and frames. So the great majority of persons using the eight or ten-frame Langstroth hive keep but one size of frame. But with the large hive used by the author of this book, shallow supers, with a 6-inch side bar, have been found preferable. They are also better liked by ladies who keep bees on account of the ease of manipulation. But it cannot be denied that a single size of frame is of value in an apiary, since combs of honey from the upper story may

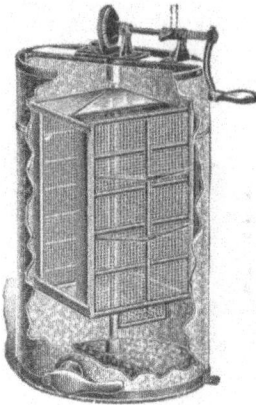

Fig. 114—A Modern Honey-Extractor Showing Inside Basket.

be given below, when the bees need feed. The shallow supers have the advantage of being occupied more readily by the bees. Nothing however needs prevent an apiarist from using both kinds.

Extracting-supers, after the first season, will be filled with comb, since the honey is removed without damaging the comb. For that reason the bees ascend in these supers much more readily, and produce a great deal more honey than they do in sections which are new, since the filled ones have been removed entirely after each crop.

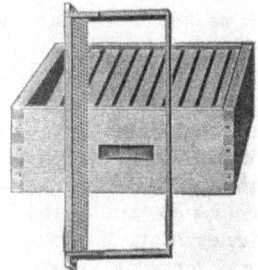

Fig. 115.—Shallow Super With Frames Instead of Sections.

The Honey Must Be Ripe

§ 163. In the production of extracted honey, the entire sealing of the combs is of less importance than in the production of comb-honey. The important requirement is that the honey be thoroughly ripened (233) before it is removed from the hive. Fresh honey usually runs like water (50), often containing as much as 75 percent of water. It takes a week or more to ripen it. This is done by the bees sending a strong draft or ventilation through the hive, night and day, during the crop, by the fanning of their wings. The novice

Major Von Hruschka, Inventor of the Honey-Extractor.

will readily see this by taking notice of the bees at the entrance on a warm evening. The file of bees which are occupied in fanning the hive runs all through their home. In this way they not only evaporate the surplus water out of the nectar, but they also keep the temperature down within reasonable limits. We have already mentioned this in the chapter on "Food of Bees" (50-53).

For the reason above mentioned we can give a colony of bees a greater amount of surplus space, when running the apiary for extracted honey than when producing comb-honey. So a less number of swarms (101) is the result. We should add also that, since the bees are supplied with the combs already built in the supers, colonies may gather a crop of surplus that could not have been expected to even begin in the sections, had they been compelled to build all their combs.

§ 164. To remove the extracting supers, the bee-escape (Fig. 112) may be used, in the same manner as with comb-honey production.

When To Use the Honey-Extractor

§ 165. If the apiarist is running his bees for comb-honey, the only use of the honey-extractor is to remove the honey from the

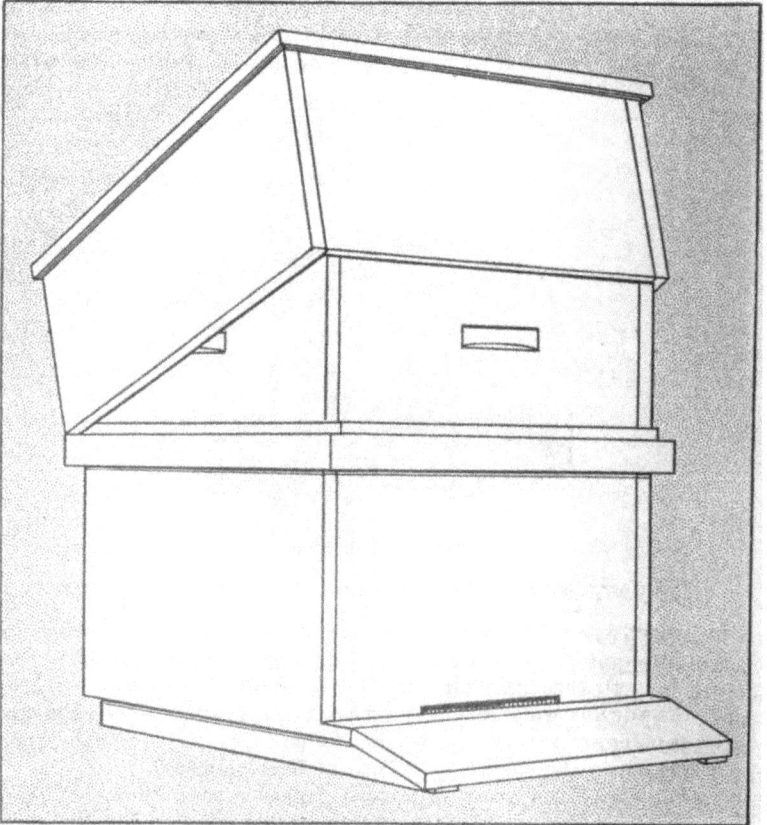

Fig. 116—The Dadant Hive. Best for extracted-honey production.

brood-combs if the breeding apartment becomes so full of honey that the queen has no more room to lay. It is also used at the end of the crop to remove the honey from partly filled sections which are unsalable. A holder for small sections is made (Fig. 118), which enables the operator to extract the honey without letting the sections down to the bottom of the wire cage of the machine.

When we are running an apiary for **extracted honey,** it is best to keep a sufficient supply of supers to store one entire crop. As soon as the first crop is over, or when it begins to decrease, if the honey is sufficiently ripe it should all be removed. But in some extraordinary seasons we have seen so large an amount of honey that it was out of the question to wait till the end of the crop. If the combs are

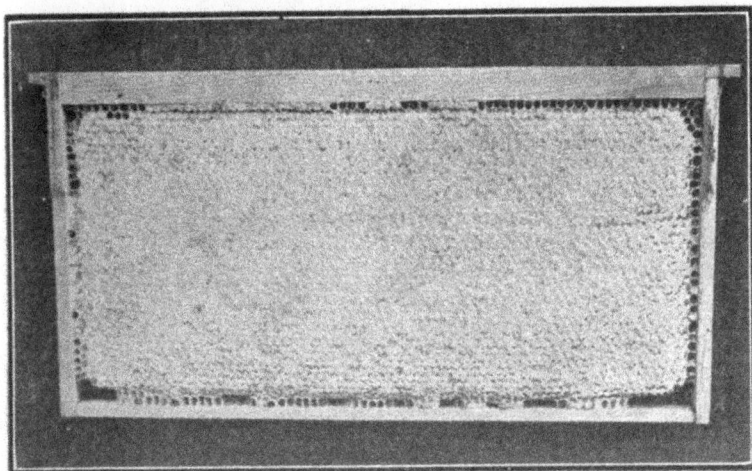

Fig. 117—Well-Sealed Honey.

sealed and the honey very thick it may be extracted at any time.

§ 166. By all means the honey from the first crop should be extracted, whether much or little, before the beginning of the fall crop, as the honey of these two crops differs much in color as well as taste, as we have said before (157).

Inexperienced beekeepers are sometimes tempted to extract too closely, and thus ruin the colony. The extractor should not be used to remove honey from the brood-chamber unless there is no room for the queen to lay, and this only during the breeding season. In the fall it is best to keep the hive-body as full as possible. The use of the extractor should then be confined to the upper story or supers.

§ 167. Since the combs must be turned over after extracting the honey on one side, in order to get it from both sides, extractors have been invented which reverse their baskets automatically while in motion.

Such an extractor is shown in Fig. 120. As the speed is arrested after extracting one side of the combs, the baskets swing over to the right or left,. according to the direction given, and the opposite side of the combs is extracted without lifting them out.

Uncapping Knife

§ 168. The combs must be uncapped before extracting. For this purpose the Bingham uncapping knife is generally used. (Fig 121).

Fig. 119—A Power Extractor

It is made of the best steel, strong at the bend near the handle. Both edges are sharp and are beveled on the side that comes in contact with the combs. This prevents the knife from adhering to the combs and tearing them, while shaving off the cappings. As both edges are alike it admits of being used for right or left hand work; the sharp point also allows it to be used in corners or uneven places. Its bevel prevents the cappings from sticking back to the comb, and causes them to drop in the uncapping can or other receiving strainer.

§ 169. A steam-heated knife has been devised (Fig. 122) which is very serviceable in cool weather, as the thickened honey is liable to stick to the blade of a cold knife.

§ 170. In very large apiaries the honey-extractor is run by the use of a gasoline engine or an electric motor. But most apiarists employ a helper who runs the machine and takes care of the honey. Where a motor is used, a honey pump is also needed.

How To Extract

§ 171. After removing the super containing the honey from the hive, carry it into the extracting-room. The honey-extractor should be carefully fastened down on some sort of platform high enough to place a bucket or other receptacle under the faucet. Uncap the combs on both sides and place them in the basket, putting combs of about equal weight opposite each other to prevent a swaying motion. A few turns of the crank will throw the honey out. Reverse the frames and extract the other side. If the weather is suitable for honey and the crop still continuing you may at once replace the combs in the hive from which they were taken, or better still, take them to the next hive to be extracted; perform the same operation using the frames just "extracted" from, to fill the places of those taken from the hive, and repeat the operation till all the

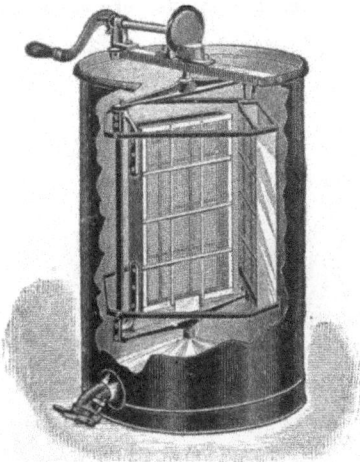

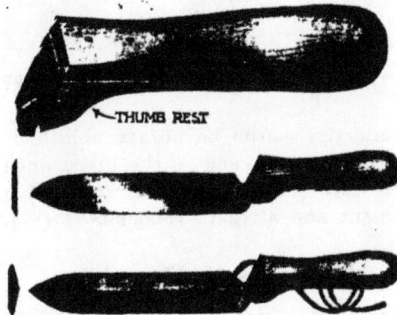

Fig. 120—Cowan Rapid Reversible Honey-Extractor.

Fig. 121—Bingham Uncapping-Knife.
Fig. 122—Steam-heated Uncapping Knife.

hives are treated in the same manner that have a surplus of honey.

By this plan, much work is saved, each colony is handled but once, the bees are less disturbed and will resume work much sooner. The frames from the last hive may be given to the first, after being emptied of the honey, instead of empty frames—if no extra combs are at hand for that purpose. This is an additional reason why only one style of hive should be used in an apiary—so that the frames may all be interchangeable. In "dividing," too, this is very essential.

Fig. 123—Original Dadant Capping Can After 37 Years Use.

However, should the season be unfavorable and the crop at end, the combs should not be returned to the hives until evening, as the honey with which they are more or less smeared by the extracting will cause excitement and attract "robber-bees (84)."

The Honey Cappings.

§ 172. The cappings which have been cut off or shaved from the combs are gathered in a large strainer, called an "uncapping can" Fig. 124). After draining the honey out of them, they may be

washed and melted. The washings are usually sweet cnough to make
honey-vinegar (241); the wax should be rendered (143) and makes
a superior article.

Tanks for Honey

In very large apiaries, where there is a possibility of unripe
honey being extracted, large tanks are provided to ripen it, by ex-

Fig. 124—Dadant Uncapping Can.

posing it to heat. In California, where it never rains in summer,
such tanks are kept out of doors till the honey is sufficiently evaporated (Fig. 125). In the central States a honey tank is best kept in the
honey house (233), never in a basement or cellar.

Fig. 125—Mendleson's Honey Tanks in California.

Fig. 126—One of Mendleson's Apiaries.

Wintering and Feeding Bees

§ 173. In Northern countries bees are wintered outdoors, and also in cellars or underground repositories. In localities soutn of the 40th degree, it is best to winter them entirely outdoors, because the proportion of warm days is great enough to allow them to take flight from time to time, and cellars, as a rule, are too warm on continued warm days. But in locations situated about the 42d degree, or north of this, it is best to put the bees away in good cellars. Between these two latitudes, the beekeeper must take his choice of indoor or outdoor wintering, according to the facilities that he has at his command.

This advice is given as applying to the eastern portion of the United States, the Mississippi Valley in particular. On the Pacific coast, or in Europe, the climate conditions differ so much that it is impossible for us to establish a rule based upon the latitude. As a rule, the beekeeper will not err if he tries outdoor wintering, wherever the bees can have a flight once a month during the winter, even though very cold days in succession may intervene. But in countries where the ground is covered with constant snows for 2 to 4 months and where the outside temperature rarely rises to the thawing point, in the shade, during the winter, cellar-wintering is best.

Cellar Wintering

§ 174. All the best authorities of the present day give the following as requisites for safe wintering:

1. An even temperature ranging from 40° to 45°.
2. Complete expulsion or absorption of moisture from the hive.
3. Perfect freedom from outward disturbances.
4. Protracted isolation from atmospheric changes in spring.
5. Exclusion of light.
6. Sufficient healthy (186) stores for winter consumption.

It is generally admitted that with these six contingencies provided for, there will be no hazard in wintering.

Make certain, before winter has come, that the colonies are all provided with at least 25 to 30 pounds of good stores, for this essential, which we have placed as sixth, should really be considered

first. It is a **sine qua non** condition of life for your bees. Should
they be short, they must be fed, before the advent of cold weather.
See the chapter on feeding which follows (186). Very weak or worthless colonies should be united together or to stronger ones (196, 115).

We must next be assured that the cellar is provided with sufficient ventilation to allow the escape from it of noxious gases and
heat generated by the bees. It is wisdom to provide a means of
letting in cold air from the outside. Although when unoccupied the

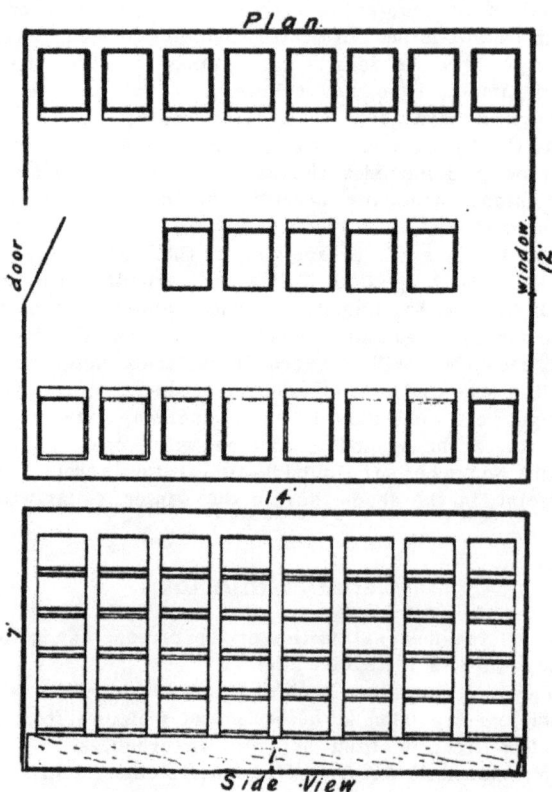

Fig. 127--Plan of Bee Cellar.　　100 hives in a space 12 x 14.

cellar may be at a mean temperature of 40° F., if 100 colonies be
placed in it they soon generate sufficient animal heat to run the
mercury up to 50°, or even more. The bee apartment should be

separate, if possible, and not so situated as to be subject to constant invasion by individuals or vermin.

§ 175. The covers should be removed from the hives, if possible, one or two thicknesses of woolen or cotton cloth—or old carpets—spread over the frames, two inch-square sticks laid crosswise of the hive, and the next one set on top and treated the same way, proceeding thus till all are neatly and carefully piled up. This work should not be done till fall is so far advanced that the bees will exhibit but little activity, when slightly disturbed. Of course, too much care cannot be exercised to do all your work gently, and if you can do so without the bees knowing they are being moved, it will be much better.

The hives may be piled up in the cellar without giving them any upper ventilation. Many persons take them in with the cover, or with a honey-board on top of the frames. In that case a large amount of ventilation is needed at the bottom of each hive. Doctor Miller, in his "Fifty Years Among the Bees" describes the bottom-board that he uses which allows a 2-inch space under the frames in the cellar. During the summer this is filled with a wooden rack

Fig. 128—A Special Bee Cellar.

which prevents the bees from building combs in this empty space.

§ 176. When all are nicely piled away, darken every nook and crack, so that should the bees venture to the entrance of the hives, they might think it a perpetual night. At least every fortnight enter your bee apartment with a dark lantern, and satisfy yourself that all progresses favorably. If the thermometer indicates above 45° F., admit cold air at night; if below 40° F., partly close the escape, to bring the mercury up to the desired temperature.

§ 177. Experienced apiarists, like Dr. C. C. Miller, assert that darkness is not indispensable. It is not, provided the proper temperature be maintained. But there are very few cellars where the variations will not make the bees restless at times, and more bees will be lost if they have light.

§ 178. Good ventilation is necessary. With pure air the bees will be much quieter than without it. Large apiaries, of say, 300 to 500 colonies may be safely wintered in a single cellar provided the ventilation be ample and in such cases it requires a large amount. But a few colonies may be easily wintered in some corner of a fruit or vegetable cellar, if only the cellar may be kept quiet and cool. Mice are very objectionable as they disturb the bees. But we have seen a dozen colonies very safely wintered on shelves above a bin of vegetables, with a tarpaulin or an old carpet as the only screen between them and the possible disturbances caused by the daily visits of the housekeeper in quest of supplies.

Fig. 129—Hives Under the Snow.

§ 179. The stowing away of the bees should be done at the first cold spell of winter, following a warm day.

To remove them, in spring, from the cellar, select a warm day in March, when the first soft maple trees open their blossoms. Take

them out of the cellar early in the day. If you have left the cover
of each hive on the summer stand, and you have your hives num-
bered, it will not be difficult to replace them in the exact position
they occupied before winter. Some say that it does not make any
difference, but the writer knows positively of two instances when
some of the bees remembered their location of the previous fall,
and went back to it. However, if you wish to put your hives in a
new location, this is really the best time to do it, especially if you
wish to move them only a few feet away from where they stood
previously (69), as most of them will have forgotten the old location.

Fig. 130—Hives Completely Covered with Snow.

Wintering bees in houses is not safe unless the wall of the house
is made so thick as to be proof against the changes of temperature.
In a room where the thermometer often rises to 60 or 70 degrees the
bees will become restless even if kept in the dark, and many of
them will perish. But if they have an opening for flight on warm
days, so that they may return to the hive, a bee-house is an ideal
location.

Wintering Bees in Clamps

§ 180. Mr. M. Quinby favored wintering bees by burying, which
is practiced by some at the present day. The mode is to dig a trench
in a hillside or ground with sufficient slope to insure drainage. This
is partly filled with straw, on which the hives are placed; boards
are slanted up in front; wooden tubes placed in position to ventilate
the pit; straw thrown on the hives, over which boards are laid length-
wise; and dirt piled over all to turn off the water.

Special Cellars

§ 181. In northern countries, where the cold days last from the
beginning of November until May, cellar wintering is indispensable.
Large honey producers have special bee cellars (Fig. 128) or caves,

Fig. 131—The Dadant Straw Mat.

well-drained and well ventilated, with a vestibule at the outer door.
That colonies may be kept in the cellar safely during a very
long winter has been demonstrated in the Province of Quebec. Mr.
Verret of Charlesbourg who winters a large apiary regularly in his
house cellar has kept his bees as long as 186 days, or from November

Outdoor Wintering

§ 182. For outdoor wintering as much honey is needed (174) as
for indoor, or perhaps a little more. The bees will consume more
honey to keep warm, and will breed earlier than if kept in the cellar, but they will also be strong earlier in the season. if there has
been no loss.

§ 183. Shelter your hives well from the wind on the north (63)
and west sides. Place over the brood-combs an empty super which
you will fill with a sack full of leaves or chaff. We use a straw mat
over the frames, and over this place forest leaves. If you have but a
few hives, and have old woolen carpets, cut these of proper size to

Fig. 132—G. C. Greiner's Double-Walled Hives Made of Dry-
Goods Boxes.

cover the frames in several thicknesses, and put them under the
cover. There must be a sufficient ventilation from below, and
moisture absorbents above, without loss of heat. Sheds and house-
apiaries, which most good beekeepers dislike in summer on account
of the discomfort of working within, are a benefit in winter. If
the bees can be placed in a shed which is kept entirely open in
front during the summer, and closed in winter, except on warm days,
they have an ideal place to winter bees, especially if it faces south.
Some hives are made with double walls and chaff packing between

the two (Fig 132).　　These are expensive, but give good satisfaction.

Winter cases, which telescope over the hives, are also very serviceable. The advantage of these is that they may be removed when spring comes so as to let the rays of the sun shine directly upon the hive-body.

Fig. 133—Apiary Packed by the Dadant Method.

An excellent method of wintering in winter-cases has received a new impetus because of being recommended by E. F. Phillips, apiarist in charge at the Department of Agriculture at Washington. This expert, with the help of Mr. Geo. S. Demuth, has made a large number of experiments in wintering and on the temperature of the cluster. A detail of these is beyond the limits of this work. But the wintering of bees in chaff hives or large winter cases has proven highly advisable.　(Fig. 132, 134, 135.)

Our own method, which is less expensive and less cumbersome and has given us good results, though perhaps a less positive success, consists in wrapping the hives with dry forest leaves, after having filled the caps or covers with leaves also. The wrapping used to hold the leaves against the hives is wire-netting. The front of the hive is left open (Figs. 133, 136).

§ 184. In all cases of packing on the summer stand, a passageway should be made by laying half-inch slats over the tops' of the frames, to afford the bees a passage from comb to comb, to reach their stores without going to the extreme ends of the frames to pass around.

§ 185. For wintering on summer stands, all preparations should be made early enough in the fall to admit of ample feeding in case of a scarcity of stores, as they cannot often be fed afterward without great disturbance (174).

Fig. 134—Greiner's Winter Case for Bees.

Feeding Bees

§ 186. Feeding bees in the fall is necessary, not only when they are short of stores, but also when the stores which they have are of an unhealthy nature (55). Honeydew, already mentioned by us in the chapter on "Food of Bees (52)" is a most unhealthy food (224) for winter. So are fruit juices, sometimes gathered by the bees in the fall when there is a shortage in the honey crop.

In localities where the bees are likely to be confined to the hive for many successive weeks, whether in the cellar or on the summer stand, all unhealthy food should be extracted and a sufficient amount of good honey or sugar syrup supplied in its place.

§ 187. Spring feeding is advisable to stimulate breeding, for we must remember that the colonies spend a great deal of honey to rear brood in preparation for the harvest. If the colony be well supplied at the end of winter, the queen will fill the cells with eggs. But often after fruit bloom, especially if the spring is backward and the weather rainy, the colonies may need additional food and a little urging to reach the honey crop without decreasing the rearing

Fig. 135—The Same Ready to Receive the Bees.

of young bees. We must bear in mind that it takes 21 days to hatch the perfect worker (37) from the day the egg is laid and that about 15 days more will elapse before the young worker becomes a field bee (36). So there must be a large amount of breeding done before the 36 days preceding the honey crop. Otherwise, should the crop be short. many of the hatched bees might find themselves consumers instead of producers.

What to Feed

§ 188. Extracted honey of your own crop, or granulated sugar reduced to the consistency of honey, is best for feeding, in the absence of good sealed honey. The poor grades of sugar and glucose are totally unfit for feeding bees. To stimulate in the spring, one-half of a pound per day is sufficient for a colony.

§189. Foreign or unknown honey should never be fed to bees,

as it may contain the germs of foulbrood (216, 222), while apparently nice and sound.

§ 190. Good honey is considered as much more suitable food for bees, for the rearing of brood, in the spring, than the very best of sugar syrup. Being their natural food, it very probably contains the necessary elements for the development of the growing insect, during its metamorphosis, for honey has been shown to contain besides

Fig. 136—The Dadant Method of Wintering.

the saccharine matter, more or less pollen, essential oils, tannin, different salts and phosphates, manganese, sulphur, iron, etc. These substances, which might be injurious if in too great proportion in the winter food, are most likely beneficial in the rearing of brood.

118

Feeders

§ 191. All sorts of feeders are manufactured. The division-board feeder (Fig. 137) which is hung in the hive in place of one of the frames is very practical, the only danger being of bees drowning in it in their eagerness to reach the food. To prevent this, a light strip is allowed to float over the liquid. It is also well to have a bunch of grass or hay so placed in it that the bees may crawl up on it. If easy means of access are given they will clean up all the feed readily.

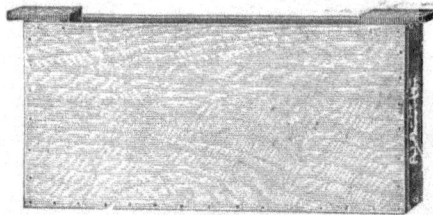

Fig. 137—Division-Board Bee-Feeder.

The Miller feeder (Fig. 138) is used in the super. Other feeders are placed in the bottom-board.

We deprecate the use of entrance feeders, as the robbers are often attracted by them and weak colonies that are fed are sometimes overpowered. We also object to outdoor feeding, for we not only may feed strong colonies which are already wealthy in stores and which take the lion's share but we may also feed strange bees. Charity begins at home. Outdoor feeding also excites the bees and causes robbing (86).

Fig. 138—Miller Feeder.

A very good and inexpensive feeder is made by using an ordinary fruit tin-can. The top is entirely removed, the can is filled with syrup and a piece of muslin is tied over the top. The can is temporarily inverted on a dish across two cleats. At first the syrup runs freely, but the atmospheric pressure prevents its continual leaking and if it is given to the bees over the brood-combs near the center of the brood-nest, it is in the easiest place of access for them. A honey-board may be used with an opening in which the can will fit exactly, so as not to allow of any loss of heat. We have used as many as five of these feeders at one time, on a populous colony in seasons of great scarcity, for winter supply. After a few days the cans are removed, and it will be found that the bees have even gnawed holes in the muslin and cleaned the cans entirely.

§ 192. To make syrup for winter feed use twenty pounds of sugar for each gallon of water. Mix the sugar with the water at the boiling point. Add a little honey (four or five pounds) if you have any which you know to be healthy.

§ 193. For spring feeding, as the bees need watery food for breeding, you may make the syrup thinner. Serve it warm, especially in spring.

§ 194. Home-made sugar-candy, commonly known as "fudge," is very good, to sustain colonies which have been neglected, to be given them at a time when the weather is too cold for them to store syrup. This "fudge" is made by heating about four parts of sugar with one part of water until it becomes thick enough. Stir it to prevent burning. When thick enough pour it on sheets of light paper. Give it to the bees over the brood-combs. It may be fed even to cellar-wintered bees, and they cluster on it as they would on combs of honey.

Helping Weak Colonies

§ 195. When in early spring you find colonies that are weak in numbers, from winter losses, they may readily be helped (if not worthless and if they have a good queen) by giving them a comb of brood from a stronger colony, after all danger of the brood being chilled has passed. On the other hand, strong colonies may be employed to cleanse out the combs containing dried-up bees that have died during the winter on mouldy combs. If your strong colonies have their hives already filled with frames of brood, then exchange combs to accomplish the purpose, but where a colony is feeble, and it is desired to build up rapidly, no disagreeable work should be imposed upon the bees to perform, for it will tax their energies sufficiently to provide pollen, water, and do the feeding and nursing necessary for successful brood-rearing. A strong colony will accomplish in a few hours that which would embarrass a weak colony for nearly a whole season.

Uniting Worthless Colonies

§ 196. Colonies that are worthless, from queenlessness or from having a poor queen, should be united, especially in the fall, when we are preparing the bees for winter. As a rule a colony which does not contain at least two quarts of bees, is unfit for wintering. However, in cellar wintering, a small colony may be more safely brought through than in outdoor wintering. Weak colonies may be united after smoking them well, by removing the combs with adhering bees and placing them together in one hive, spraying them with peppermint water by an atomizer to give them all the same scent. Give them ventilation and reduce the entrance till sunset, placing them where the stronger of the two colonies stood. The poorer one of the queens should be removed.

Put a slanting board in front of the hive, which will cause the bees to mark their home anew (69). On the third day remove the board from the front. No hive should occupy the old stand, from which the queen and bees were removed, for several days.

Another very practical method, recommended by Dr. Miller, is to put the weaker of the two hives to be united over the stronger one, with a sheet of paper between the two hives. The bees will gnaw the paper and will slowly become acquainted, so their reunion will take place usually without fighting. If the weaker colony occupies only two combs, these may be lifted out bodily with the bees and inserted in a space already prepared in the other hive, with the same use of a sheet of paper between.

Bee Pasturage

§ 197. As civilization clears away· the forests and upturns the prairie sods, it may increase or decrease the possibilities of the honey crop. When we remove the honey locusts and the basswood trees in clearing our forests, when we plow up the lowlands of the fertile valleys and destroy myriads of summer blossoms to replace them with fields of corn or small grain, we decrease the prospects of honey yield. But when our modern pastures become filled with white clover which has followed in the wake of civilization, when

Fig. 139—Cherry Blossoms.

we plant alsike clover, alfalfa, sweet clover or melilot, fruit trees of different sorts, black locust (robinia), etc., we establish a new flora for honey production.

It is well for the beekeeper to become acquainted with the principal honey plants of the region in which he lives and to increase the honey flora in all practical ways.

Trees For Shade and Honey

§ 198. Every home can be beautified by a judicious selection of

ornamental shade-trees, and where the roads, streets and lanes are
nicely bordered with them, the market value of the property will be
increased greatly. For this purpose the basswood or linden (Tilia

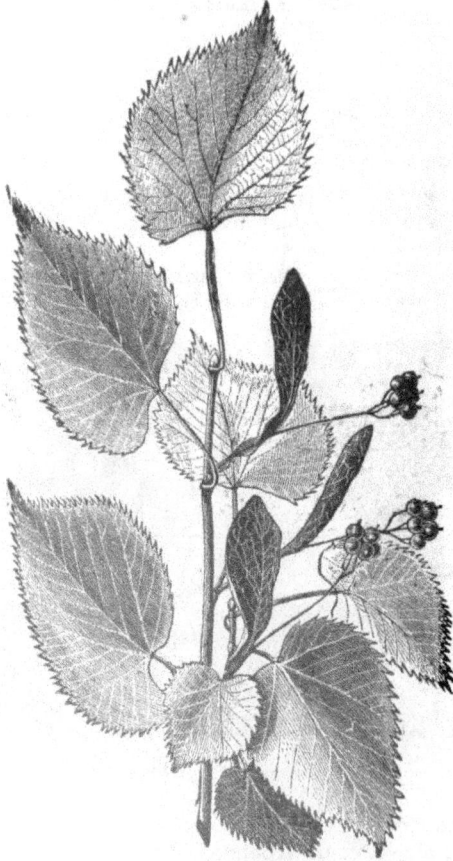

Fig. 140—Basswood or Linden Leaf and Blossoms.

Americana) is one of the most desirable. Its rank, thrifty growth,
large, glossy-green leaves, perfumed flowers, adaptability to almost
any soil and climate, make it one of the most desirable for lawn or
lane. It can be transplanted with certainty. It blooms in early
July, and yields a white, aromatic honey, of good quality.

§ 199. The tulip tree (Liriodendron tulipifera), often called poplar, is also of rapid growth, hardy, and easily cultivated.

§ 200. There are two or three varieties of willows, all good honey-
producers, which are adapted to all sections of our country. The

Fig. 141—A row of Basswood or Linden Trees in McHenry Co,. Ill.

little care required to propagate them is a recommendation in their
favor, especially in moist soils.

§ 201. The black locust or Robinia (pseudo-acacia) are almost
certain honey-producers. Although the duration of bloom is but
limited, they yield a supply of rich nectar, and bees will literally
swarm among the highly-perfumed blossoms. G. W. Demaree. of
Kentucky, wrote as follows regarding the locust:

"The time of year in which it blooms nearly filling the interval
between the late fruit-bloom and the white clover, makes it an exceedingly valuable auxiliary to the honey harvest in the Middle
States, if not elsewhere. It is a most profuse honey-bearer, rivaling
the famous linden in quality, and only inferior to the product of the
latter in color. Locust honey cannot be said to be dark in color. It
is of rich pale-red color, when liquid; but when in the shape of comb-
honey, its appearance, if removed from the hive when first finished.
is but little inferior to that of superior clover honey. It becomes exceedingly thick, if left with the bees till the cells are thoroughly

sealed, and its keeping qualities are therefore most excellent. The
trees are planted by the side of fences, in waste places, and on poor,
worn out lands. They may be propagated from the seeds, or by trans-
planting the young trees from one to three years old. If the ground
is plowed in the spring, and the locust seeds planted on the hills with
corn, or with other hill-crops, and cultivated the first year, the
young trees will grow with great rapidity, even on very poor lands."

Fig. 142—Tulip or Poplar Leaf and Blossom.

§ 202. Fruit-trees of all kinds are eagerly visited by the bees,
and yield pollen as well as honey.

Plants for Field and Roadside

§ 203. When the apiarist is so situated that a few acres of land
can be devoted to bee-pasture, we would advise that such selections

be made with a view to answering the double purpose of producing honey, and grain or winter forage for stock. Although convinced that a profit may be realized from land devoted to honey-producing alone, if a remunerative profit can be obtained from its cultivation for honey, and other returns be derived from the crop, it is an additional net profit, less the cost of harvesting and marketing.

§ 204. There are, however, many bee-keepers whose grounds are very limited, but in whose immediate vicinity are lanes and alleys but little used, or waste commons and worn-out fields, which, with

Fig. 143—Black Locust or Robinia.

little labor and less expense, could be made profitable to an apiary thus becoming spots of beauty and sources of revenue, instead of remaining evidences of sloth and a public reproach.

§ 205. White clover (Trifolium repens) is too well known to require particular description, and is associated with too many pleasant recollections to call for commendation. The lawn would indeed, seem incomplete, without the clover carpet with its velvet surface of mingling white and green, giving out its ambrosial perfume while the bees in myriads sing from flower to flower. White clover will always be a welcome tenant of waste corners, nooks, and roadsides, and no farmer need be told of its value for pasturage. Its honey is not excelled by any other.

It is the main honey plant of the east and middle west in the United States and millions of pounds of its white honey are sold. As its growth is volunteer, it does not need further recommendation.

§ 206. For field or commons our next preference is decidedly given to sweet clover or melilot (Melilotus alba). Being one of the hardiest plants we have it will withstand any degree of winter's cold or summer's heat, and its deep-penetrating and wide-spreading roots, admirably adapt it to any variety of soil, whether wet or dry, sand or clay, loam or gravel. Being remarkably thrifty in growth, it is superior to red clover for soiling, and can be successfully grown in locations where the latter will prove a failure. But its greatest recommendation· for the general bee-keeper is

Fig. 144—White Clover in Bloom.

Fig. 145—White Clover After Bloom.

the fact that it requires no especial cultivation, thus making it particularly desirable for roadsides and commons.

When sown for forage, it is planted in the fall with winter wheat, or in early spring it can be sown with wheat, oats, or rye, without detriment to the grain.

As the sweet clover seed has a very hard coat, it will sometimes delay in its germination until another season, if the seed is quite dry when sown. The modern way is to scarify the seed after hulling it, by passing it through a machine which scratches its coat and renders it more permeable to moisture. Seed thus treated produced 95 per cent growth in three or four days.

Sweet clover blooms and yields nectar continuously in its second year, from June till frost. It is a biennial and dies at the end of the second season. To secure good hay from it, it should be cut

Fig. 146—Part of a 22-Acre Field of White Clover in Iowa.

when the stems are about two feet in height, about two inches above ground, as it will not grow again if cut too close to the ground. Three crops of this hay may be harvested in the same season. The cutting of it before bloom retards its honey production, but this is only delayed beyond the blooming time of white clover, which is the main honey plant of the Mississippi valley. Horses and cattle which are unacquainted with the taste of sweet clover often hesitate to eat it. But when once they learn to eat it they prefer it to other hay.

Many people used to consider sweet clover as a noxious weed; the prejudice against it has been overcome when it was ascertained that it does not spread to cultivated fields and is easily controlled.

It smothers and readly destroys very noxious weeds, such as the ragweed.

A noted apiarist wrote of it: "I suppose I owe my wonderful summer success largely to the sweet clover. We had the hottest and driest season we ever had—no rain from June 15th until Septem-

Fig. 147—Sweet Clover Just Before Bloom.

ber 15th. The hotter and drier the more honey, seemingly. Sweet clover, as a weed! Although it has been growing in our roads, on waste land, along railroads, and on our hillsides for twenty-five years, it does not seem to get into the fields, except where water has carried the seeds into low places."

W. T. Stewart, of Kentucky, says: "Melilot is best sown in the fall, but will grow any time or anywhere, except on a flat rock."

To sum up, it is worth more to the farmer for soiling than red clover, because of its thrifty growth; it is a more reliable pasture for cattle, sheep, etc., than red clover, because it will thrive on soils where red clover sickens; it will yield much more fodder than red clover, because it will stand two or three cuttings; and it lacks but seven per cent of possessing the nutritious properties of red clover. We can add, we believe it is worth the cost of cultivation to the bee-keeper, for honey alone, even though he is not the possessor of a four-footed animal.

§ 207. Alsike or Swedish clover (Trifolium hybridum) is also a good grazing and honey plant, and sown in connection with dairying

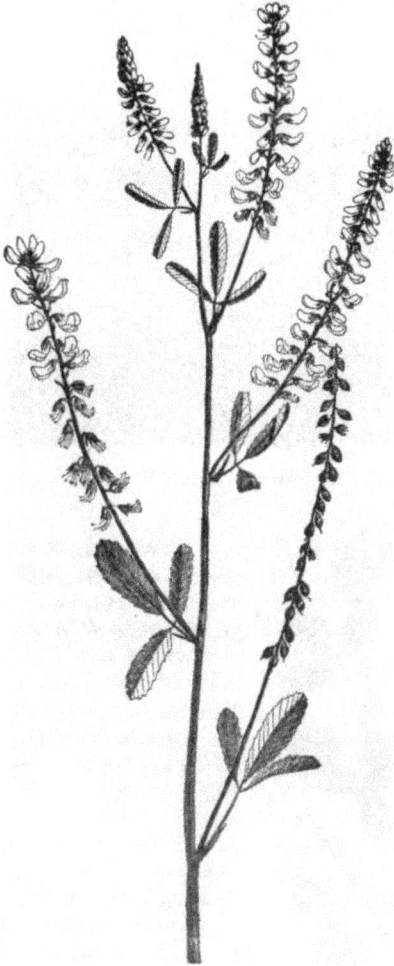

Fig. 148—White Sweet Clover (Melilotus alba) in Bloom.

pursuits of stock-raising, will prove doubly valuable. Mr. M. M. Baldridge, of Illinois, who has devoted much careful study to this clover, says:

Fig. 149—Sweet Clover in South Dakota.

Fig. 150—Alsike Clover.

"The stem and branches are finer and less woody than the common red, and when cut and cured for hay, it is perfectly free from fuzz and dust. It does not turn black, but remains the color of well-cured timothy. The bees have no trouble in finding the honey, as the blossoms are short, and the heads no larger than those of white clover. The blossoms at first are white, but soon change to a beautiful pink, and emit considerable fragrance. It is not advisable to cut this clover more than once each season, but it may be pastured moderately during the fall. When sowed by itself, four pounds of seed is sufficient for an acre; but this is not the best plan to pursue, especially on dry western prairie land. It is much the best to mix it with timothy or common red clover, or both. When thus mixed they are a help to each other, and two pounds of alsike seed to the acre are sufficient. Alsike clover as a fertilizer must be as good a plant as red clover, as the roots penetrate much deeper and are more numerous (Fig. 151). It is a clover which every farmer can and should cultivate, whether he keeps bees or not, as it is superior to the common red for hay or pasture for all kinds of stock."

§ 208. One of the best honey-plants that the modern methods have brought to the front in the United States in the past thirty years is alfalfa or lucerne. This plant, which has been grown in

Fig. 151—Alsike Clover Root and Crown average size, one year old. Red Clover Root and Crown, one year old.

Europe for centuries as forage for horses or cattle, is one of the best honey-producers in all the irrigated valleys of the arid or semi-arid States. The honey of Colorado, gathered from this plant alone, is shipped in dozens of carloads. Three crops or more are harvested.

In some localities it does not yield honey. Such is the case in some sections of Illinois.

§ 209. There are several varieties of the mustard (Sinapis)

which furnish honey. These have been extensively cultivated for the seeds alone, and always have a commercial value. The length of season for bloom is quite extended, and where a dearth of honey-pasturage prevails, bees work on them vigorously. They bloom during July and August.

Fig. 152—Milkweed—a Honey-Plant Whose Pollen Sticks to the Feet of Bees.

§ 210. Buckwheat (Fagopyrum esculentum) is familiar to every Northern beekeeper. Its grain always commands ready sale in market, and the honey, though dark and strong is prized for manufacturing purposes. It furnishes an excellent winter food for the bees, and when well-ripened will enable the producer to avail himself of all the white grades of honey stored earlier in the season. In

early morning the bees work on the buckwheat with great enthusiasm, and gather honey from it rapidly; but during the middle
and latter part of the day they entirely neglect it, unless the weather
be quite cloudy and humid. In the Southern states, we have been
told, buckwheat is worthless as a honey-producer, and, in fact, the
same is true of many localities in the Middle and Northern States;
but where it does produce honey abundantly, it is well worth cultivation.

Weeds as Honey Producers

§ 211. In addition to the plants above-named which are useful

Fig. 153—Alfalfa or Lucerne.   Fig. 154—Buckwheat in Bloom.

as crops as well as for honey, there are excellent honey-producing
plants which have been classed as noxious weeds because they grow
voluntarily in cultivated fields or stubble. Among them are a
number of persicarias, of the polygonum family, commonly called
by the names of knotweed, heartsease, smartweed, etc. Most of these
are good honey producers, blooming from the end of July until frost.

A large-flowered bidens (bur marigold, beggar-ticks, spanish
needles) producing yellow blossoms, in lowlands, around marshes,
a few days before frost, produces very fine golden-yellow honey of

good quality. Both these and the knotweeds are annual.

Among the perennial weeds which produce much honey are to be ranked the asters, blooming also in autumn, and of which there are many different kinds with blossoms white, purple, and blue. They are called by some "wire weeds" on account of the toughness of their stem. In the East, the golden-rod, in the West the sage, are excellent honey producers.

Fig. 155—Buckwheat Field.

There are hundreds of other plants which yield both honey and pollen, but they are unimportant and description of them is out of the limits of this work. They will be found mentioned in the Langstroth-Dadant book "The Hive and Honey Bee."

Fig. 156—Heartsease or Persicaria.

Observation Hives

§ 212. No man can keep bees successfully unless he becomes well acquainted with the habits of bees. This he may acquire to a certain extent by reading. But there is nothing like practice to to learn a thing well. Therefore, we should try to study the habits of the bees from the bees themselves. For this an observation hive is needed.

Fig. 157.—Observation Hive inside of Sitting-Room Window.

A good observation hive (Fig. 157), is composed of only one comb in a frame with glass on both sides. This is supplied with either doors or a black cloth cover. The doors are better excluders of light, but the opening and closing of them often jars the bees slightly and disturbs them.

The hive may be placed in a window with an entrance at the front so the bees may go to the fields and supply their needs. We usually stock up an observation hive in the spring by taking a good comb of brood from one of our best colonies with plenty of bees to keep the brood warm. The first thing they do is to rear a queen. We thus witness the different changes through which the brood passes, the hatching of the queen, the bringing in of pollen, honey, etc. It is an endless source of amusement and instruction. At the end of the season, as it is difficult to winter bees in so small a hive, it is advisable to unite it with some populous colony and re-stock it in spring, with bees and brood.

Fig. 158—Frank C. Pellett, Associate Editor of American Bee Journal at His Observation Hive.

Enemies of Bees

§ 213. The **enemies** of bees are not numerous. A few birds, among which we shall name the king bird, eat bees. But their

Fig. 159—The European Death's-Head Moth Natural Size.

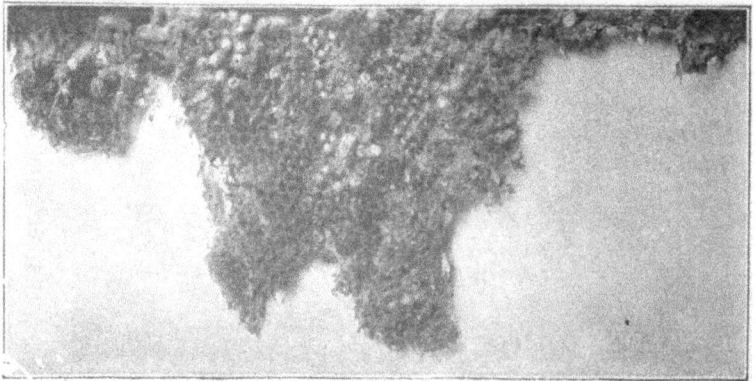

Fig. 160—Brood-Comb Destroyed by Moths.

damages are so insignificant that they are hardly worthy of mention.
Ants sometimes make their nest over the bee-hive, to take advantage of the warmth of the bees. They may be readily driven

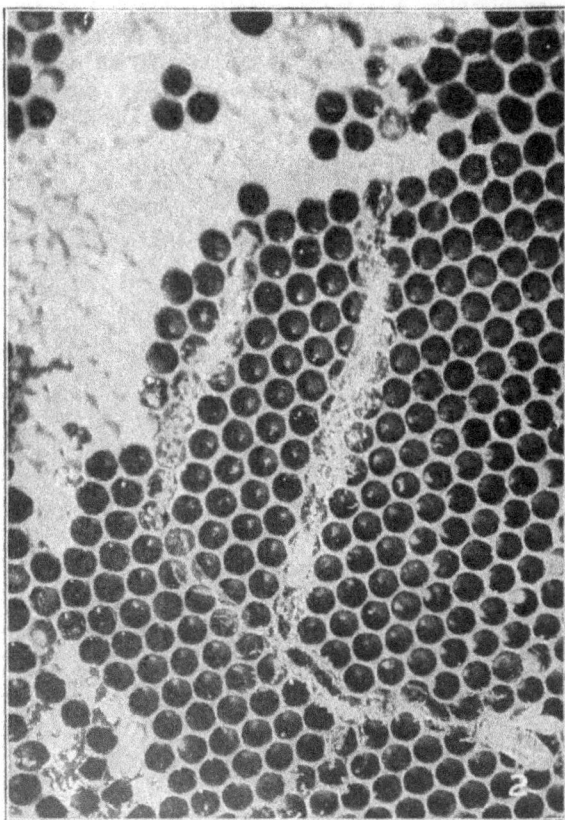

Fig. 161—The Web of the Moth-Larva.

away by placing salt or powdered sulphur where they congregate.
The bee-louse or "braula-cœca," and the death's-head moth which
enters the hive to feed on the honey, exist in Europe but are unknown
here.

The Beemoth

§ 214. The most active enemy of bees is the beemoth, which lays

its eggs in neglected combs, especially in old combs. The larvæ hatch
and devour everything in their reach, making webs or galleries (Fig.
161), through the combs. Colonies that have more combs than they
can cover, or queenless colonies, especially in the fall, at a time when
the moths have already reared two broods and are therefore numerous, are often rendered worthless by the ravages of the larvæ of the
beemoth. Two different kinds of moths are known, but the larger or
"tinea melonella" is the principal depredator. Luckily they cannot

Fig. 162—Lincoln Monument at Springfield, Ill., Reproduced in
Beeswax.

stand the winter in cold rooms where the temperature goes below
zero. It is only when accidentally sustained in some corner of a
populous colony or in combs in a warm room that the moths can
reproduce from one year to another.

§ 215. For a remedy there is but one rule—keep your colonies
strong, and they will destroy the moth. Do not keep any combs in
exposed places. When moths are discovered in combs they may be

readily destroyed by the use of brimstone or bi-sulphide of carbon. The latter ingredient should be used with care, as it is inflammable. Spread a little on a rag and place it over the combs, shutting down the box in which they are contained.

Burning brimstone in a beehouse will destroy the moths if it kills the flies (227).

Diseases of Bees and Treatment

§ 216. Of all the diseases of bees the most dreaded is foulbrood. Foulbrood attacks the larva, which dies in the cells. It is infectious and must be treated with promptness and care. There are two different varieties of this disease, American and European foulbrood. The names do not indicate that either kind originated in the countries named but only that the descriptions of each have originally been given in those countries. They are due to bacteria in the form of a bacillus which in the case of American foulbrood is carried in the honey (189) mainly, while in the other disease, it seems to be transmitted otherwise and perhaps at times through the queen, though the exact mode of infection is only conjectured. For their detection and cure, we cannot do better than quote the State Inspector of Wisconsin, Mr. N. E. France, who has probably had more experience with these diseases than any other man in the United States:

American Foulbrood

Symptoms

§ 217. "(1.) Brood in combs badly scattered, many empty cells, cappings dark and sunken, some with holes in cappings, part of the brood hatching while others are dead: The dead larvæ of a dark brown color, or blackish according to age. The lightest colored will upon inserting a tooth-pick draw out much like stale glue when warm.

"(2.) Dried Scales. If the disease has reached advanced stages, all of the above conditions will be easily seen. According to its age or development there will be either the shapeless mass of dark brown matter on the lower side-wall of the cell, or the dried scale.

This scale is nearly black and dried hard to lower side-wall of comb, and as thin as side-wall of the cell."

McEvoy or Starvation Treatment

§ 218. "In the honey season, when the bees are gathering honey freely, remove the combs in the evening and shake the bees into their own hives; give them frames with comb foundation starters and let them build comb for four days. The bees will make the starters into comb during the four days and store the diseased honey in them, which they took with them from the old comb. Then in the evening of the fourth day take out the new combs and give them comb foundation (full sheets) to work out, and the cure will be complete. By this method of treatment all the diseased honey is removed from the bees before the full sheets of foundation are worked out. All the foulbrood combs must be burned or carefully

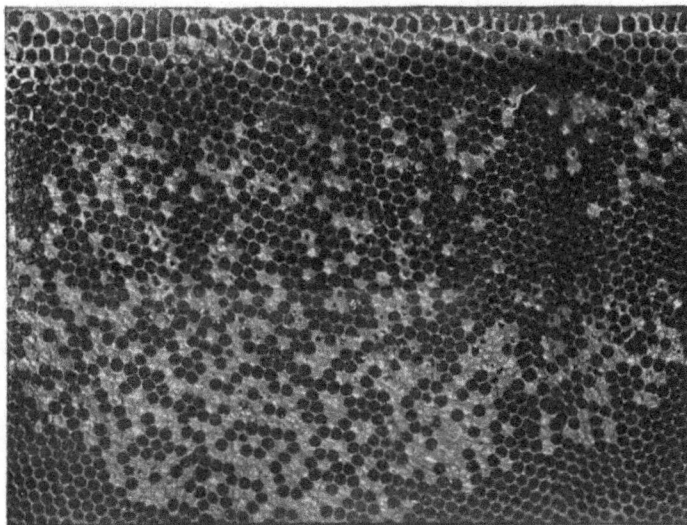

Fig. 163—American Foulbrood—Part of a Brood-Comb.

made into wax after they are removed from the hives, all the new combs made out of the starters during the four days must be made into beeswax, on account of the diseased honey that would be stored in them. The curing or treating of diseased colonies should be done in the evening, so as not to have any robbing done, or cause any of the bees from diseased colonies to mix and go with the bees of healthy colonies. By doing all the work in the evening it gives the bees a chance to settle down nicely before morning, and there is no confusion or trouble.

"Sometimes the bees, deprived of all their combs and brood, will on the following day leave their hives, and may enter several others, and may carry the disease with them. To avoid this, it is well to give the bees some feed, either honey from perfectly healthy colonies, or, if any doubt, then feed syrup of equal parts white sugar and water, giving the feed in the upper hive above the bees, so as to prevent robbing.

"Honey from infected hives should never be given to bees, un-. less it is boiled for fifteen minutes. Then it may be used safely. Such boiled honey will be very dark colored, and bees do not like it.

"Never let bees get to infected honey; better bury it deep in the earth.

"This treatment is most reliable, and has been tested for many years in all climates. I find the greatest number of failures where the operator is not careful in treating. Ever remember that a

N. E. FRANCE
Fig. 164—A Practical Expert on Bee Diseases

single drop of infected honey, or piece of infected honey, or piece of infected comb, carelessly left exposed, will be enough to give the disease to as many colonies as come in contact with it. I am unable to find proof that such honey is injurious to persons eating it, but find plenty of evidence that it will kill larvæ or young honey-bees.

"Hives well scraped, are safe to use again, and if the frames are boiled under boiling water for some time, they are also safe to use again. Comb foundation from infected wax will be safe to use, as

I have proven in 60 cases in Wisconsin. Queen-cages may contain disease, so, to be safe, I remove the queen into a new cage before introducing, and place old cage and attendants in the fire. If queens are from known healthy colonies, they can be introduced in the shipping cage in which they arrive.

Fig. 165—Mr. France Holding a Comb of Foulbrood to Show the Proper Angle to Detect the Scales on the Lower Side of the Cells.

Avoid all bees robbing infected or just treated hives."

Instead of boiling, which is a slow process, the hives and fixtures may be singed by the flame of a tinner's blow-torch, or by smearing with coal-oil and applying a match, extinguishing the flames after a few seconds.

The odor of American foulbrood is similar to that of a carpenter's glue-pot.

European Foulbrood or Black-brood

§ 219. In 1898 to 1900 this fatal disease destroyed many profitable apiaries in New York State, until State bee-inspectors were appointed with instructions to do all possible to abate the disease. After many experiments they have succeeded in curing it almost everytime.

Symptoms

§ 220. "No ropy or stringy dead brood; no marked foul odor; not attached to the comb.

"The young brood soon after hatching from the egg into the larval stage turns yellowish in color, sometimes quite dark along the back line. The head end of larva becomes pointed, standing out from walls of the cell. The body continues to dry, the skin toughens, and remains in the cell loose, **never adhering** to the walls of the comb as does American foulbrood. There is a sour odor sometimes, in the early stages, much like sour apple pomace. The colony soon weakens, giving robber bees and the waxmoths access to the remains. The moths eat wax only, not the infected bees in the cell."

Treatment

§221. Make the hive queenless, by killing or removing the queen. Allow all the brood to hatch, which will usually take place in about 21 days. Then give the bees a new queen reared in a healthy colony. Sometimes caging the queen for 10 to 21 days is sufficient. The bees cleanse the cells of diseased brood and the colony may overcome the disease. But it is best to change the queen, as such queens given to healthy colonies have been known to carry the disease with them. Italian bees overcome the disease much better than the common bee. It is said that Carniolan and Caucasian bees are also superior to the common bees in this respect. However each of those races has been known to be affected, so that they are not entirely immune. Much is to be learnt yet concerning foulbrood diseases.

If you are afraid to treat the bees without help, secure the address of your State Inspector of Apiaries, and write him. Most of the States are now prepared to help fight the disease by official means. It is an offense punishable by fine to allow foulbrood to exist without treating it. There is no doubt that it can be destroyed by the above-mentioned methods.

Pickled Brood or Sacbrood (White)

§ 222. Pickled-brood is a similar disease to the one above mentioned (European foulbrood), but it is of a mild nature.

In all the above-named diseases nothing needs be destroyed except the combs containing the dead American foulbrood, which is so ropy and sticky that it can never be cleaned out by the bees. But the honey is unsafe for the bees to use , and should not be returned to them. It is for that reason that honey which you do not know should never be fed to bees (189).

Combs of diseased colonies containing no dead brood may be

rendered (143) into beeswax. The boiling destroys the germs of the disease.

May Disease

§ 223. There is a disease of the adult bees, which is variously called May disease, paralysis, constipation, etc., and often ceases after a few days. But occasionally colonies that are affected lose so many bees that they become worthless. Infrequently, the queen herself may contract the disease and die. The bee's abdomen becomes distended, the insect is apparently in great misery, and crawls about as if partly paralyzed. Italian beekeepers, who have had considerable experience with it, have recommended feeding the colony with honey or syrup strongly saturated with tonics, such as essence of rosemary, lavender, ginger, etc. Powdered sulphur blown over the bees and at the entrance is said to stop the infection. But powdered sulphur kills the brood and must be used with caution. Some apiarists hold that the disease is caused by excessive dampness and that it may be stopped by shaking the colony on dry combs.

A very similar disease, which has caused great havoc in the British Isles, is known as Isle-Of-Wight disease. In the United States it has done but little damage.

Diarrhea

§ 224. Bee diarrhea in the latter part of winter and early spring is a malady that affects some apiaries. The bees discharge watery excrements over the hives and combs, producing a dark appearance and offensive odor. The cause is either fermented honey, improper food, long confinement, or too warm and poorly-ventilated quarters.

Fruit-juices, harvested during a dearth of honey, are the most frequent causes of diarrhea. These juices, insufficiently sweet to keep from fermenting (186), are stored in the combs like honey. They should be extracted and replaced with good honey or syrup. Honeydew from plant-lice is also a cause of diarrhea when cold weather confines the bees to the hive a long time.

Usually, diarrhea disappears with the first flights of the bees. But in a protracted winter it is difficult to cure. It is much more easily avoided by pure food, given at the opening of winter, than stopped after it has once begun. When syrup is fed, none but the best granulated sugar should be used. Commercial glucose is death to the bees, and in most cases they refuse to accept it.

Marketing of Honey

§ 225. The marketing of honey is a subject that interests every apiarist. Today, comb-honey is the preference for table use, and if we would cater to the public want, we must produce that article in the most attractive shape.

Fig. 166—A 3-ton Load of Honey.

Assort and Grade the Honey

§ 226. All honey should be graded, and a scale of prices be established. An apiarist, compelled by his needs, may sell honey at the commencement of the season for any price offered, and thus unintentionally break down the market, by giving a start at too low a rate. Systematic organization could and should help this state of affairs. The Colorado Honey Producers' Association has taken the lead in this matter and has organized a corporation which handles many carloads of honey each season, at remunerative prices. They use a uniform double-tier shipping case containing 24 sections of honey.

Grading rules have been adopted by the National Beekeepers' Association at different times as also by the above-mentioned Colo-

Fig. 167—A Honey Delivery Wagon for Retailing.

rado firm. In addition to this the National Government now requires that each section of honey be marked as to its minimum weight. This is to prevent deception.

Management of Comb-honey

§ 227. Comb-honey should be taken from the hive as soon as it is finished, or as soon thereafter as possible (156). Mr. G. M. Doolittle wrote:

"No apiarist can expect to have his honey sell for the highest market price, if he allows it to stay in the hives for weeks after it has been sealed over, allowing the bees to give the comb a dirty yellow color. by constantly traveling over it. All comb-honey producers know that there always will be cells next to the section that are partly filled with honey but not sealed over, and when taken from the hive, if the section is turned over sidewise, the honey will run out, making sticky work. The remedy for this is a small, warm room. Bees evaporate their honey by heat, and therefore, if we expect to keep our honey in good condition for market, we must keep it as the bees do, in such a position that it will grow thicker, instead of thinner all the while. Our honey-room is situated on the south side of our shop, and is about 7 feet square, by 9 feet high. We have a

Fig. 168—Getting a Crop of Comb-honey Ready for Market.

large window in it, and the whole south side is painted a dark color to draw the heat. In it the mercury stands from 80° to 90°, while our honey is in it; and when we case it for market, we can tip our sections as much as we please and no honey will drip, neither will any of the combs have a watery appearance—all will be bright, dry and clean. But if we keep honey thus warm, the moth (215) will make its appearance, and make it unfit for market, by gnawing off the sealing from the beautiful combs.

"We build a platform on either side of our honey-room, of scantling, about 16 inches high, and on this we place the sections so that the fumes from burning sulphur can enter each one; in about two weeks we fumigate, by burning ¾ of a pound of sulphur for every 200 cubic feet in the room. We take coals from the stove and put them in an old kettle, so as not to get anything on fire; pour on the sulphur and push it under the pile of honey, and shut up the room. Watch through the window, and in 15 minutes after the last fly or bee that chances to be in the room has died, open the door and let out the smoke, for if it stands too long, the smoke may settle on the combs and give them a greenish hue. As there may be a few eggs that have not yet hatched, we fumigate again in about ten days, after which the honey will be free from moths, if you do not let millers into the room."

Honey in the comb is a luxury—a fancy article—and our first care should be to produce it in such a manner as to command a fancy price. It must captivate the eye of the consumer, and tempt him to purchase Comb-honey should be put up in uniform cases, and not veneered, i. e., the combs inside should be just as good as those in the exterior of the case.

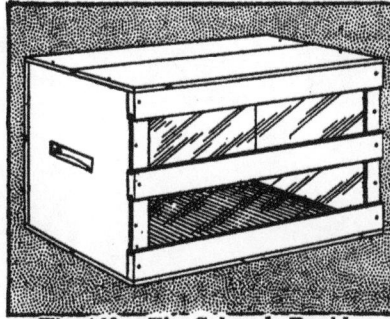

Fig. 169.—The Colorado Double Tier Case.

Shipping-cases for Honey

§ 228. Cases in which to pack comb-honey for shipment Fig. 169-170 are made in various sizes, holding from 12 sections to 28 in a single tier.

Fig. 170—No-Drip Shipping-Case.

The most satisfactory shipping-case has inside a folded paper pan at the bottom, upon which are tacked small strips crosswise whereon to set the sections. This forms what is known as the "no-drip" shipping-case (Fig. 170). Should there be any dripping of the honey it is caught in the paper pan, and the cross strips hold up the sections so that they do not rest in the honey-drippings. Of course only perfect sections of honey should be packed for market, and not any that are at all in a leaky condition.

Shipping Comb-honey to Market

§ 229. A few directions on packing comb-honey for shipping by railroad may be useful. It is best to have a large crate holding perhaps 16 of the 12-section cases, or 8 of the 24-section cases. First put about four inches of straw in the bottom of the large crate, then place in the cases of honey, not forgetting to put straw at the sides and ends of the large crate, as it is filled with the cases of honey.

Fig. 171—Well-Sealed Honey in Sections.

The straw acts as a cushion. After nailing on the top pieces enclosing the large crate, nail a three or four inch board on each side, about a third or quarter of the way down from the top, to be used as handles for carrying the whole crate of perhaps 200 pounds of honey. Thus two men can carry it easily, and there is practically no danger of breakage, if properly packed (Fig. 173).

Management of Extracted Honey

§ 230. The marketing of extracted honey is an important matter, for a good article, attractively put up, will always command the best

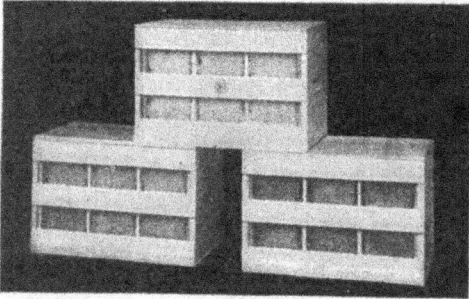

Fig. 172—Shipping-Cases Filled with Comb Honey.

Fig. 173—Crate for Shipping Cases of Comb-Honey.

price and it is therefore, of the utmost importance to producers to have honey put up in the best shape.

Every beekeeper should fully supply his own locality, and he should let it be distinctly understood that it is the pure honey taken from the combs by centrifugal force—that nothing is added to it, and nothing taken from it but the comb.

It should also be kept before consumers that granulated honey can be reduced to its liquid state by placing the honey in a vessel of warm water. When thus liquefied, it so remains for some time before again granulating.

§ 231. Consumers may be sure of a wholesome article by purchasing granulated honey and reducing it. If heated above 160 degrees there is danger of spoiling the color and ruining the flavor. Honey contains the most delicate of all flavors—that of the flowers from which it is taken. A good way is to set the vessel containing the honey inside another vessel containing hot water, not allowing the bottom of the one to rest directly on the bottom of the other, but putting a bit of wood or something of the kind between. Let it stand on the stove, but do not let the water boil. It may take a day or longer to melt the honey. If the honey is set directly on the reservoir of a cook-stove,or on a steam or hot-water radiator, it will be all right in due time.

Honey which is heated darkens in color, unless kept at as low a temperature as possible during the process and cooling it promptly when once melted.

If the product is for a home market, the producer must study the local preference regarding the size and style of package, as well as the grade of honey most easily disposed of. As far as practicable, keep each grade of honey separate; it is a mistake to suppose a few pounds of inferior or different shade honey will make no difference in a large bulk of white clover honey, or that thereby a better rate will be obtained for the second-grade article. Instead, the result will most likely be to class it all as second-grade, and the price of all will be depreciated.

Where to Keep Honey

§ 232. As honey has strong deliquescent tendencies, or in other words absorbs moisture readily, it is advisable to keep it in a warm, dry place, whether it is comb or extracted honey. In such a place it will gain in quality, while in a damp cellar or basement it will gather moisture and readily sour. A well ventilated honey house (227), or lacking this, an attic, or other dry spot, is suitable.

Ripening Honey

§ 233. The nectar gathered from the flowers cannot be called

honey until the evaporation or ripening process (50) has so far
gone on that the bees have commenced capping it over. If it be extracted (163) before it is capped by the bees, as some apiarists recommend, on account of the quantity being thereby greatly augmented,
then it should be ripened before it is placed in tight packages or
shipped, or it is liable to ferment and sour.

§ 234. Tanks for ripening extracted honey (172) are usually made
of galvanized iron or tin and of different sizes. For a small apiary
an extractor can (Fig. 123) is sometimes used. The honey is drawn
off into retailing vessels before it begins to granulate.

Tin Pails for Honey

§ 235. For retail packages, tin pails (Fig. 174), with close-fitting
covers, are the best. Purchase by the gross or in lots of 1,000 or
more, the price is inconsiderable.

Fig. 174—Straight Pails for Holding Honey.

A neatly printed label should be gummed or pasted on each pail,
stating the amount and kind of honey, name of apiarist who put it up,
and giving in a foot-note directions for liquefying the honey in case
it granulates.

Tapering pails are heavier and stronger than the straight pails;
the covers are deeper and the top-edge of the pail is doubled over.
A smaller size is also made to hold about one pound.

Glass Containers

§ 236. If smaller packages are wanted, then use glass jars, or
tumblers. Jars and tumblers, like the tin pails, should be tastefully
labeled.

Large Packages

§ 237. The square tin cans (Fig 175), furnish excellent packages for safely shipping extracted honey. Each can holds about 60 pounds,

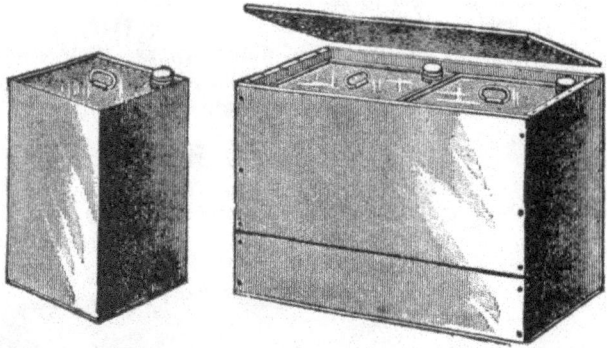

Fig. 175—Square Cans for Shipping Extracted Honey.

and two of them may be shipped together in one crate or box (Fig. 175). There is no leakage in transit, if even moderately well handled. A slat one inch square should be placed over each can, before nailing the cover down, in order to protect the screw cap.

Honey as a Food

§238. The use of honey instead of sugar for almost every kind of cooking, is as pleasant for the palate as it is healthy for the stomach. In preparing blackberry, raspberry or strawberry short cake, it is infinitely superior.

Pure honey should always be freely used in every family— Honey eaten upon wheat bread is very beneficial to health.

Well ripened honey has the quality of preserving, for a long time in a fresh state, anything that may be laid in it or mixed with it, and to prevent its corrupting, in a far superior manner to sugar; thus many species of fruit may be preserved by being laid in honey, and by this means will retain a pleasant taste and give to the stomach a healthy tone.

In fact, honey may replace sugar as an ingredient in the cooking of almost any article of food—and at the same time greatly add to its relish.

Mr. Alin Caillas, a French chemist and analyst of note and author of a booklet entitled "The Treasures in a Drop of Honey," serving as lieutenant in the French army at the front in 1915, reported to us that a number of the soldiers in his company having become debilitated by the constant use of dry canned foods and salt meat were brought back to health and vigor in a very few days, by the daily use of a regular amount of honey which he had had the good luck to purchase from an apiarist living at a short distance behind the battle lines and the trenches.

Give Children Honey

Prof. Cook says: "We all know how children long for candy. This longing voices a need, and is another evidence of the necessity of of sugar in our diet. * * * Children should be given all the honey at each meal-time that they will eat. It is safer, will largely do away with the inordinate longing for candy and other sweets; and in lessening the desire will doubtless diminish the amount of cane-sugar eaten."

Ask the average child whether he will have honey alone on his bread or butter alone, and almost invariably he will promptly answer, "Honey." Yet seldom are the needs or the tastes of the child properly consulted. The old man craves fat meat. The child loathes it. He wants sweet, not fat. He delights to eat honey; it is a wholesome food for him, and is not expensive. Why should he not have it?

Honey Best to Sweeten Hot Drinks

§ 239. Sugar is much used in hot drinks, as in coffee and tea. The substitution to a mild-flavored honey in such uses may be a very profitable thing for the health. Indeed, it should be better for the health if the only hot drink were what is called in Germany "honey-tea"—a cup of hot water with one or two tablespoonfuls of extracted honey. The attainment of great age has in some cases been attributed largely to the life-long use of honey-tea.

Is Honey a Luxury?

§ 240. While it may not be a necessity, no more it is a luxury than is butter or beefsteak. Some writers have pointed out that because one could not live on honey alone, it was a luxury and should be sold as such. One could as well live on honey alone as on butter alone, yet no one regards butter as a luxury.

Fig. 176—Four Articles of Equal Food Value—7 Ounces Honey, One Quart Milk, 15 Ounces Codfish and Ten Eggs

A fair basis of values of food products is the actual food units which they contain. In order to secure reliable information as to the food values of the products which we wish to compare with honey, we have taken the table compiled by Hon. W. B. Barney, of the Iowa food and dairy department. With this table at hand we went to a retail store where the usual retail prices prevail, and purchased different products of equal food value.

Fig. 176 shows 3 articles, with food value equal to 7 ounces of honey. For the quart of milk we paid 10 cents, for the codfish 20 cents, and for the eggs 25 cents. Milk and eggs are generally recognized as necessities, yet as far as food value is concerned the eggs cost more than twice as much as the honey, and the milk is slightly higher in price.

In Fig. 177 is shown a 12-ounce steak which costs at retail 15 cents, yet which, according to Mr. Barney's table, is only equal to 7 ounces of honey in food value. When beefsteak is regarded as a necessity even by those who are working for the lowest wages, why should the impression grow that honey is a luxury at half the price? Nine cents worth of cream cheese is equal to 7 ounces of honey, yet even this costs more than the product of the hive. Thirteen cents worth of walnuts are necessary to equal the small jar of honey. Since extracted honey usually sells at less than 16 cents per pound at retail, 7 cents will not be far from the cost.

Fig. 177—Seven Ounces of Honey is Equal in Food Value to 12 Ounces
Round Beefsteak, 5.6 Ounces Cream Cheese or
8½ Ounces Walnuts.

Fig. 182 shows that 8 oranges, which cost 20 cents, supply an amount of food equal to 7 ounces of honey, and 5 bananas, at 25 cents per dozen, cost 10 cents.

The following table shows the amount of the various items required to supply food value equal to 7 ounces of honey, according to t.e above mentioned authority. The retail prices that prevail at th:s time are also given:

Honey, 7 ounces 7c	Boneless codfish, 15 ozs 20c
Cream cheese, 5.6 ounces... 9c	Oranges, 8 ....20c
Eggs, 10 ....25c	Bananas, 5 ....10c
Round beefsteak, 12 ozs.....15c	Walnuts, 8½ ounces 13c

The above items are in general use, and few if any of them are regarded as luxuries. By reference to the above table it will be seen that as far as actual value is concerned, honey is one of the cheapest of the ready-prepared foods. Only such raw products as potatoes, corn meal, beans, etc., which must be prepared for the table after purchase, are cheaper in food value, at current prices, than is honey.

Fig. 178—7 Ounces Honey, 5 Bananas and 8 Oranges—Honey is One of the Cheapest Foods in the Market for Actual Nutritive Value.

Honey Vinegar

§ 241. We have explained (172) that the cappings of the honey combs, after draining the honey from them, may be washed and the sweetened water used to make vinegar. Honey vinegar is superior to any other kind and may be almost colorless. The sweetened water from washing the cappings should be tested if used for vinegar. If no saccharometer is at hand, just put an egg into the liquid. If the egg does not show at the surface, add more honey, mixing it well. If it floats and shows more than a spot the size of a dime at the surface, add more water. If you wish to make honey vinegar and have no sweetened water, mix one and a half pounds of honey with each gallon of hot water. When it is at a temperature of between 70° and 90° add some ferment, fruit juice, grapes preferred, or yeast, to start the fermentation. If the liquid is kept in a barrel, do not fill it over two-thirds, as air is needed. The vessel should not be closed, but simply covered with a cloth or a very fine screen to keep out insects. The first fermentation will be alcoholic. But the air will soon supply the germs of acetic fermentation if the temperature is kept above 70°. In a few weeks or a few months at most, you will have excellent vinegar. Honey-dew or dark unsaleable honey may be used for this purpose.

THE END.

List of Illustrations

The numbers refer to the Figure numbers.

Index

The numbers refer to the paragraph and not to the page.

www.ingramcontent.com/pod-product-compliance
Lightning Source LLC
Chambersburg PA
CBHW021431180326
41458CB00001B/227